TECHNOLOGY—THE KEY TO BETTER ENVIRONMENT

Books by Donald C. Blaisdell

Technology—the Key to Better Environment
Government and Agriculture
American Democracy Under Pressure
International Organization

Technology—the Key to Better Environment

Values, Profits, and Growth
in Post-Industrial Society

Donald C. Blaisdell

An Exposition-University Book
EXPOSITION PRESS   **NEW YORK**

FIRST EDITION

SBN 0-682-47713-3

LIBRARY OF CONGRESS CATALOG CARD NUMBER: 73-77582

Manufactured in the United States of America

Published simultaneously in Canada By Transcanada Books

Contents

Foreword

In a chat with a friend three years ago I observed how shameful it was that we in the U.S.A. had allowed our environment to become so polluted. My friend, a retired utility executive, agreed, but added that he thought the electric power service industry was making good progress in clearing up pollution and that the country as a whole was inclined to try to go too fast in repairing the pollution caused by a century and more of rapid economic growth. Out of our exchange came a decision to prepare a paper or an article together, he to expound environmental policy from the point of view of the electric power service industry, I to argue that our ways of making environmental policy were historically deficient in not allowing sufficient place for the public, as distinct from private, interests.

This book is the result of that decision. Unfortunately, my friend was unable to make his contribution. So I have to try to cover both the general policy toward our surroundings and the specific case of the electric power service industry.

But if he was responsible for none of the composition, he has, nonetheless, been helpful in supplying me with references and materials. To him, too, I owe a debt of gratitude for some of the thoughts in the last chapter.

Acknowledgements are due, too, to my brother, Thomas C. Blaisdell, of the University of California, Berkeley, to friends like Mrs. Annabel Jannsen, George C. Lincoln, U.S. Army (ret.), Mrs. Gwynne Smith, and Mrs. Karel Van Zonneveld, all of whom have in one way or another made helpful suggestions.

For the way the material is presented and for the suggested program for accelerating present and developing new technology for an improved environment, I am, of course, entirely responsible.

—Donald C. Blaisdell

Preface

The American people like to think of themselves as a resource-ful people. Having devised and operated a Constitution which has encouraged and promoted the settlement of a domain of continen-tal, even extracontinental proportions; having invented and devised machines, tools, and manufacturing processes which are the envy of the world; having become accustomed, over two hundred years, to a growth-oriented economy able to satisfy expanding consumer demands with ever-decreasing demands on labor; yet the American people in the 1970s indulge increasingly in self-doubts. One reason is the question many ask themselves: Is America destined to con-tinue on its historic course of using more and more of its own and the world's resources in the inefficient, wasteful, and environmen-tally-degrading ways which are by-products of our growth-oriented system of private economic enterprise? Or can we consciously redirect our use of science-based technology so that adaptations of that system needed for stopping environmental degradation can be adopted? It is the purpose of this book to explore how the American people have, through their government, tried to come to terms with science-based technology as it affects environment.

1.
Environmental Policy and the American Ethos

For about three-quarters of the two centuries since the Declaration of Independence the American people had no environmental policy, except possibly by default. Nor did their governments, federal and state, unless the tacit invitation contained in the Constitution to exploit a continent can be interpreted as such.

No doubt it is inexact and unscientific to speak of *the* American ethos, as if the spirit which has inspired our attitudes toward morals, practices, and beliefs is the same in the twentieth century as that of three hundred, two hundred, or even a hundred years ago.

Yet, despite changes, which are inherent, some threads in the American ethos can be traced from the present to the past, a connection which neither space scientists nor members of the counterculture (youth culture) can break.

Biologically we are creatures of planet earth; man cannot adapt himself to the space environment.[1]

Culturally we are, similarly, creatures of the past with a heritage which each generation receives, more or less, from its forebears both enriched and at the same time impoverished by the cultural environment in which it is reared.

Central to the ethos of America is the myth of the inexhaustibility of resources. As regards air and water our forefathers may be excused for such self-delusion. Even as regards nonrenewable resources it is still not hard to understand. Bemused by the frontier psychology, Americans for seven generations and more have

---

[1]Rene Dubos, *So Human an Animal*. New York: Charles Scribner's Sons, 1968.

followed their destiny ever westward. Thus considered, the war in Southeast Asia is only the latest in a long series of frontiers creeping westward across North America, the Pacific, to Asia. Nor is it accidental that the same president under whom American military were first committed to Vietnam made the decision to land Americans on the moon during the sixties. The frontier psychology in America has been individualistic and growth-oriented. At the same time it is a greedy psychology, chewing up resources with little reference to future needs or, indeed, to the effects of exploitation on people regardless of generation.

In the American ethos foresight—some thought of the future—was not entirely lacking. The insurance industry illustrates it. But even here the foresight shown is selective. Insurance has been available against certain but not against all risks. Promoters of the first post roads and canals and of the first railroads could not be guaranteed against loss from some alternative means of transport; of the first telegraph against loss from alternative electronic means of communication; of ferrous metals against loss from the development of nonferrous metals; of organic chemicals against loss from development of synthetic chemicals, etc. Nor has the insurance principle been applied to those utilities developing electric power from fossil fuels. They have not been guaranteed against loss from the application of atomic energy to electric power generation. Resource depletion has not been insurable. Here, however, the tax laws have granted favorable treatment to those exploiting nonrenewable resources. Probably the most recent illustration of the nonapplicability of the insurance principle is in regard to cigarette advertising. Tobacco manufacturers were unable to insure themselves against loss from blocking off access to radio and television. Nor were the broadcasting companies able to insure themselves against loss of cigarette advertising revenue from the tobacco manufacturers. Both in the risks to be insured against, and in the natural resources to be compensated for when nonrenewable, the spirit animating the beliefs and practices of the American people has been highly selective.

A central feature of the attitude toward exploitation of resources has been the necessity for continued growth. Moreover, for generations government finance has been postulated on a constantly rising national product; budgets have been drawn, and estimates of expenditures and of revenue have been based, on expansion of the economy. Only in the last decade or two have voices been raised against the uncritical acceptance of the idea of a constantly rising gross national product. Even today they are voices crying in the wilderness.

Moreover, in the rush to provide goods and services for a constantly increasing population some serviceable principles of governmental operation have fallen by the wayside. The old idea that carrying the mail was a government service to the people has had to give way to the modern concept of running it as a business —that is, with a balanced budget, expenditures not exceeding revenues. Also, new principles have been adopted on the assumption that all government services are quantifiable in money terms. Thus, cost-benefit as applied by Defense Secretary McNamara to his department was later extended by President Johnson to the government as a whole, a concept and formula for rationalizing decision-making regardless of the service involved. The wisdom of applying the same principle to the Post Office Department and to the Department of Defense is overlooked or ignored.

Until recently America's economic leaders have rarely if ever had room in their ethos for the concept of regulated growth of the economy based on prudent, scientific use of its resources. Neither have historical events been lacking which offered opportunities to cancel out the lack of foresight. Development of resources through free, private enterprise was motivated for the most part by a spirit and a belief which regarded public regulation in any form as anathema. Few questioned the optimistic assumption that natural resources were inexhaustible. Nor did they doubt that in value terms private was superior to public enterprise. For them the only regulation needed by the economy would be supplied by the market. And as regards the natural setting the climatic differences between the eastern seaboard and the Mississippi valley, on the one hand, and the semiarid regions west of the

100th meridian of longitude on the other, were relatively unimportant.

Events seemed to bear out the validity of the tacit assumption that old mistakes made in the exploitation of the country's resources would be more than corrected by the new industries created from new scientific advances. The new technology of mining and metallurgy; of industry and manufacturing; of capital raising; and of transportation, communication, distribution and exchange, all together wiped out memories of financial panic, of unemployment, of inflation, of scarcity of capital, of blasted hopes and seared aspirations. The genius of America was production, demonstrated for all the world to see by triumphant participation in two world wars, which required little more than half of our productive capacity for victory. Before the recent revolution in communications, in transportation, and in electric energy generation, preparation for war, in the thirties, provided the necessary stimulus to overcoming the effects of the Great Depression.

It is premature to assess even tentatively, the place to be occupied by radio and television and electronics technology in general in the future ethos of America. The judgment may be hazarded, however, that, so far these examples of the American genius have found far more use as gadgets, as playthings, and as means of entertainment than they have as means of increasing man's understanding of his place in the universe. Electronics may make it possible to place men on the moon and bring them back to earth. Similarly, electronics has introduced drastic changes in weapons technology. But in pondering the significance of these two applications of electronics; in reflecting on the place to be accorded these marvels of communication in the hierarchy of goals for America; and in shaping public policy adequately to deal with them in the future development of the American people, these require far more use to be made of the communications media to increase public understanding of their benign and threatening potentialities, and more inventiveness in incorporating their use and regulation in public policy.[2]

---

[2] See chapter 2.

HISTORICAL BACKGROUND

In the beginning the granting of patents and of copyrights by the federal government suggested a simple pattern of assimilation of technology into society and one that offended no believer in free enterprise. However, the manufacture of cannon, arms, and ammunition also by the federal government, and in government arsenals, was a portent of more complex patterns, among them some decidedly socialistic in fact if not in theory and destructive of the modern free enterprise system.

Similarly, the contract system at an early date sank deep roots into federal governmental practice, when the government in 1792 decided to use its postal powers to contract for stage lines and riders instead of establishing its own.

Also, when the federal congress established the War Department in 1789 and the Navy Department nine years later not only had the government by that time adopted a mixed pattern of public construction of yards and arsenals and of private contracting for supplies. It had also received authority to establish institutions of higher education, which it did in establishing the military and naval academies (1802 and 1845).

Also dating from 1802 was the Corps of Army Engineers, a channel through which technology in the form of internal improvements and annual rivers and harbors appropriations, from 1824, started running with increasing volume and strength.

Throughout the nineteenth century invention, discovery, science, and technology welded and shaped the form and functions of government as well as the nature of the economy. Fulton's successful steamboat ultimately required a bureau of steamboat inspection to regulate the application of steam to navigation. Lewis's and Clark's explorations and the discovery of gold in California stimulated migration, while Morse's invention of the telegraph and the construction of the transcontinental railroads provided the means by which people and resources were brought into juxtaposition.

Simultaneously, the General Land Office was established to

dispose of public land to western settlers. Originally placed in the Treasury Department, it was transferred thence to the new Department of the Interior in 1849. Reaping machines were patented in the 1830s. Not for thirty years, however, in the midst of the Civil War, was the Department of Agriculture established and authorized to undertake research in the agricultural sciences. Nor was any regulation of the railroads attempted until 1887, when the Interstate Commerce Commission was set up as an independent agency.

Meanwhile, experience with the Customs Service (in the Treasury), the Pensions Office (in the War Department), and the Patent Office (in the State Department) resulted in the bureau being adopted as the basic federal administrative unit. The bureaucratic system, with appointment of the bureau chief and financing often removed from control of the cabinet officer nominally responsible, was to assume major significance in the politics, lobbying, and pressures of assimilating science and technology into public policy in the twentieth century.

And by the 1880s, when John Wesley Powell, as chief of the Geological Survey, mounted his campaigns in Congress to tie western growth to water resources by drawing state and county lines according to river basin rather than astronomically-fixed boundaries, the later struggle between environmentalists and apostles of progress in our own time was foreshadowed.

CORPORATE INVENTION AND RESEARCH AND DEVELOPMENT

For a century and more after the adoption of the federal constitution the power of Congress to grant patents operated primarily to encourage capital to develop and market inventions and, secondarily, to reward invention by natural persons. Then, in the 1890s corporate enterprise began the practice of team research which now dominates research and development. Meanwhile, the principles of interchangeable parts and of specialization of labor were applied to the manufacture of arms, then, increasingly, to handicraft industries, until, with the assembly-line principle of the early twentieth century, and automation, the foundation stones of American industry and manufacturing were laid. Simultaneously, the connection between war and accelerated research and development became increasingly clear throughout the nineteenth and twentieth centuries, until it reached its peak in World War II in the Office of Scientific Research and Development (OSRD). Federally-subsidized research now shares the field with privately-financed research on, roughly, a fifty-fifty basis.

As a problem in public policy the environment can be traced to the early years of the twentieth century, when corporate industry took over the art of invention. While the unwise use of natural resources was already widespread by the turn of the century, particularly as regards the forests and the prairies between the Mississippi and the Rockies, only with the beginnings of organized research by corporate industry did the nationwide spread of pollution of air, water, and soil become predictably certain.

One of the biggest polluters of all—the automotive industry—was among the first to subsidize research. With the mass production of his Model T, Ford started it. Then General Motors was organized in 1907, set up a special engineering department in 1911, and changed the name to General Motors Research Corporation in 1920 when it took over the Dayton Engineering Laboratories Company.

Other pioneers in rationalizing research were General Electric and American Telephone & Telegraph. The electric power service industry, with its related manufacturing industry, is another large polluter. General Electric, one of the largest manufacturers of electrical generating equipment, was perhaps the earliest to subsidize research. Incorporated in 1892, it started its research laboratory in 1901, thus enabling it to consider itself a pioneer in inaugurating the corporate movement of financing research.

Another pioneer was AT&T. Bell Telephone Laboratories were organized in 1924, owned jointly by Western Electric and AT&T. Western Electric, AT&T wholly-owned manufacturing subsidiary, set up experimental laboratories in Chicago and New York, which were consolidated in 1907 in New York. AT&T's own department of development and research was merged in 1934 with Bell Telephone Laboratories. Thus, all development and research of the telephone monopoly were brought under one supervision, one name, and one organization.

A full exposition of the phenomenal growth of subsidizing research by corporate industry as a principal factor in sharpening the issue of environmental degradation would require similar reference to other big polluting industries, such as the tobacco manufacturing industry and the invention of the cigarette-making machine; the pulp and paper industry, of paper-making machinery; of the chemical industry and perfection of synthetic fibers; of the iron and steel and canning industries, of food canning; of frozen foods and packaging and bottling machinery by the food and beverage and packaging industries; of insecticides, herbicides, and fungicides by the oil and petrochemical industries; of detergents by the soap and detergent industries; of oil tankers and pipelines by the oil and shipbuilding industries, etc. The pharmaceutical industry would have to be examined, likewise the broadcasting and advertising industries, and the adoption of atomic energy by the electric power industry.

All such inquiries would provide further evidence, if it were needed, of the central place occupied by the invention industry, to coin a term, in bringing the environment to its present pass. But such inquiry and exposition here are unnecessary. Enough has been said to make the point: In the first half of the twentieth century the rate of growth of science-based technological research mounted by corporate industry must be regarded as a triumph of American genius. But it guaranteed, as well, that the degraded condition of the environment would confront the government, science, and the people eventually with a challenge which could not be ignored.

On the governmental side, the National Academy of Sciences, the first of the formal organizations sponsored by the federal government for encouraging research, came into being in the 1860s. It was given an operating branch—the National Research Council (NRC)—during World War I, when the Council of National Defense was also established, by statute. The legislation setting it up was still on the statute books in the early days of

World War II, and in 1940 was used as the legal basis for the National Defense Research Committee (NDRC). Later, in the spring of 1941, the NDRC, together with a new Committee on Medical Research (CMR), became the Office of Scientific Research and Development (OSRD), with an advisory council, established by an executive order of the president.

Thus, the federal government belatedly organized both research and development in the cause of American defense. OSRD was established to mobilize not only the scientific talent already employed by the federal government, but also scientists and technologists of the private sector consciously to exploit inventions already made as well as to make new inventions. Indeed, as regards development the latter was to turn out to be more important than the former. The Manhattan District, independent of OSRD, exploited through brilliant scientific technology discoveries about the atom made decades earlier, while success in the art of invention itself occurred more directly under OSRD's auspices.

Organizationally, the distinction between pure and applied science was a belated one. Not for roughly another century following the setting-up of NAS was applied science—technology and engineering—recognized organizationally with the establishment of the National Academy of Engineering (NAE).

It is unimportant, however, to try to draw a line between the organization of research in pure and that in applied science. What is important is to emphasize the coming of age under federal government auspices of both research and development as the culmination of the movement deliberately to seek, collectively and on a team basis, the secrets of nature and to exploit them for wartime purposes.

Acceptance of the idea of deliberately inventing new things by and through collective efforts does not belittle the significance of the great inventions by individuals in the nineteenth and early twentieth centuries. The stature of Priestley, of Fulton, of Morse, and of McCormick; of Bell, Edison and Marconi; of Madame Curie, of Becquerel, De Forest and others such as Fermi and Oppenheimer is not lessened by interpreting the accelerated growth of science and technology and invention in terms of team effort. Rather it is to acknowledge the greatness of their accomplishments, as individuals, and to fit them into an historical movement which, in our time, threatens evil as well as good to the human race.

It is worthwhile emphasizing the deliberate nature of research and development in modern times, as distinct from the earlier reliance on individual insight and hunch. How the creative process works undergoes more and more research as time goes by. But it seems clear even now that it is a combination of rational thought and intuition, varying from one individual to another. And, applying the old adage of two heads being better than one to the conscious search for truth did not await the determination of the exact proportion of reason and intuition in the creativity process. The invention of the art of invention, with its perception of enhanced creativity through joint efforts, may in the future be appraised as the supreme example of man's creativity.

In the United States it would appear in retrospect that one of the great stimulants to technology was the expansion of markets for durable consumer goods by extending credit for financing the purchase of automobiles. The inventor of this device is not recorded. But in the 1920s a story circulated around Morningside Heights, in New York, that the inventor was Edwin R. A. Seligman, professor of economics at Columbia University. Seligman had been retained by General Motors to study the economics of instalment selling.[3] Not included in the agreement, so the story goes, was a verbal postscript going something like this: "In making your study don't forget that we want to sell automobiles."[4] The

---

[3]Edwin R. A. Seligman, *The Economics of Instalment Selling; a Study in Consumers' Credit, with Specific Reference to the Automobile Industry.* New York and London: Harper & Brothers, 1927, 2 volumes.

[4]The author has been unable to find anyone who can corroborate his recollection of this episode. It is, perhaps, not irrelevant that Professor Seligman, in the preface to his work, states that ". . . My chief indebtedness, however, is due to Mr. John J. Raskob, vice president of General Motors Corporation, and Mr. C. C. Cooper, president of General Motors Acceptance Corporation. These gentlemen courteously put at my disposal the complete records of their respective organizations and were prodigal of

study concluded, of course, that instalment selling was a good way to sell automobiles, that is, good both for General Motors and for the country. Thus, a change in the American ethos, which only a few suspected was taking place in the 1920s, became tacitly accepted by the 1950s, when General Motors' Charles E. ("Engine Charlie," to distinguish him from General Electric's Charles E.) Wilson made his immortal pronouncement that "what's good for General Motors is good for the country."

THE CONSERVATION MOVEMENT

Simultaneously with the rationalization of the art of invention came the conservation movement. Far-sighted, if not visionary, individuals preached the gospel of prudent, instead of wasteful, use of resources. Interest groups, organized around such ideas as conservation of fauna and flora generally, came into being, as did special interest groups dedicated to the protection, enjoyment, and preservation of birds, fish and wildlife, forests, and wilderness areas. Interest groups became pressure groups, urging on Congress withdrawal of public land from settlement, establishment of national forests, reforestation. Schools of mining and forestry were established, to provide trained geologists to locate new mineral deposits and trained foresters to help the lumber industry and, at least in part, to study and teach principles of conservation and multiple use management of public lands. Theodore Roosevelt, a sickly youth who later became police commissioner of New York City, renewed his health while exploring and ranching in the South Dakota Badlands, became an ardent outdoorsman and exponent of the vigorous life, and, as twenty-sixth president of the United States (1901-9) signed into law much legislation putting a statu-

---

that assistance which comes only from ripe experience and masterful leadership. Without this generous help this work could not have been written." Two members of the Columbia University community whom the author has consulted recall the way that such consultant relationship with industry cast doubt on Professor Seligman's objectivity as a scholar, one going so far as to say that Seligman's fellow social scientists "ostracized" him as a result.

tory foundation under the conservation movement.

But the conservation movement was only one facet of the American attitude toward natural resources and a secondary facet at that. Far more important was the attitude typified by the ruthless practice of lumbermen of "cutting and getting out"; of the sheep and cattle wars against homesteaders for use of the public domain; of the iron and steel industry in dumping its waste into Lake Michigan and Lake Erie; of the chemical industry and the pulp and paper industries in using river systems as sewers. At the same time that Chase S. Osborn, "the iron hunter," was exploring northern Michigan and discovering the great iron deposits there and in Minnesota, lumbermen had already stripped northern Wisconsin of its virgin timber. At the same time John Wesley Powell, explorer of the Colorado River canyon, was synthesizing his General Plan for the settlement of the arid lands, settlers were streaming eastward from those same lands, victims of weather, unfounded optimism, "Manifest Destiny," and unwise government policies of land settlement. And even while Theodore Roosevelt, as president, was giving prestige and providing leadership for conservation, he was trying to "bust the trusts," industrial mergers grown fat on the unwise exploitation of natural resources. The preachments of conservation leaders, men like John Muir, Gifford Pinchot, John Burroughs, Luther Burbank, Henry Fairchild, even of Theodore Roosevelt himself, while expressive of the slowly growing conservation movement, made little immediate impression on a country, expanding in population, area, and outlook, which gave too little heed to their thoughts of the future.

The precise year when the country opted for unlimited rather than prudent growth cannot be pinpointed. But it could have been at a point in time not much closer to the turn of the nineteenth to the twentieth century than we are today from the twentieth to the twenty-first. At that time, when Congress refused to accept as national policy the General Plan proposed by Major John Wesley Powell for organizing the then largely unsettled West, it opted for unlimited expansion and development, regardless of resources, and rejected a plan based on relating man to his resources, particularly water, which other societies—Pueblo, New Mexico Spanish, and Mormon—were already proving viable.[5]

In its essence Powell's General Plan was simplicity itself. Let water, the essential prerequisite to settlement and development, determine the economics and government of the semi-arid West. In practical terms this meant extensive rather than intensive agriculture, ranching and farming, the latter on a prudent scale, rather than irrigation.

Similarly, organization of the government and administration of these lands should be tied to the water. The basin boundaries of the great rivers should be used as state boundaries. Watersheds of the tributaries of the rivers should be made the areas of local government.

No one knows, of course, whether adoption of Powell's Plan would have prevented the dust storms, the soil erosion, the floods, the lawsuits between states over the waters of the great interstate rivers which have occurred in the Great Plains and the Colorado River basin. What we do know is that in our own time regional river basin development is getting more and more consideration as the remedy for the ravages of the malpractices of the past.[6]

And, it should be added, many of the environmental problems now confronting the country between the Atlantic seaboard and the 100th meridian (western Kansas) would have been avoided, or at least lessened. The adoption of Powell's General Plan would have signified the triumph of a philosophy of land use based not on continuous economic expansion but on adaptation of human culture to nature, not the other way around. An echo of Powell's scale of values can be found in the writings of a contemporary bacteriologist, Rene Dubos; especially his emphasis on the high priority he assigns to the proper use of land. Using land improperly, for short-range economic benefits, we reap ecological crisis, as at present. Used properly, and with ecological understanding, land

---

[5]John Upton Terrell, *The Man Who Re-Discovered America*. A biography of John Wesley Powell, New York: Weybright and Talley, 1969.

[6]John Wesley Powell. *The Exploration of the Colorado River*. Abridged from the first edition of 1875, with an introduction by Wallace Stegner. Chicago: The University of Chicago Press, 1957, p. xiv.

remains productive and desirable for human life.[7]

The beginnings of a policy toward the environment, heralded by the conservation movement and stimulated by leaders in the scientific, academic, and youth communities, began to appear in the 1960s. Together with a remarkable growth in interest in ecology, both in its biological and sociological meanings, and a grudging acknowledgment of a deteriorating environment by a population bemused by the multiplication of electronic marvels of the fifties and sixties, successive Congresses and chief executives responded with projects designed to reduce the rate of pollution of land, water, air. Projects abounded—against pollution of the sea by pesticides and oil spillage; of the air by noise, smoke, soot, and noxious gases; of the land by pesticides and wastes; of fresh water by industrial and human wastes, detergents, and of humans by drugs. Many projects reached the statute books; sizable sums of money for implementation were authorized. Lesser appropriations were made and little legislation was permanent. In other words, if we may say that public policy is taking shape in an absorption of conservation by environmentalism and ecology, it should be added that the content of the policy has not gained clear outlines as yet. For that continued and increasing pressure from many publics will be necessary.

Probably the most important single influence was Rachel Carson's *Silent Spring,* which appeared in book form in 1962.[8] Typical of those believing thus is Irstin R. Barnes, "The Naturalist" columnist of *The Washington Post.* According to Barnes, no book in his lifetime has had the width and depth of impact on men everywhere as *Silent Spring.*[9] It is not difficult to accept this appraisal in view of the campaign mounted by the agricultural chemical industry and its allies in Congress and the federal bureaucracy to discredit both the book and its author, on the ground that the book was unscientific and based on emotion. Those who accused Miss Carson of emotionalism in presenting her facts have

---

[7] Rene Jules Dubos. "The Genius of the Place," *American Forests,* vol. 76, no. 9, September 1970, 16-19, 61-62, at 19.

[8] Rachel Carson, *Silent Spring,* Boston; Houghton Mifflin Co., 1962.

[9] "10 Springs Later," *The Washington Post,* September 24, 1972.

now, in turn, been discredited, according to Barnes. This matter is alluded to in another place (pages 28-29).

THE PRIVATE COMPONENT OF
THE PUBLIC INTEREST

Theoretically, there are many ways of thinking about the public interest: as a bundle of values; as the product or sum of particular interests; as an abstraction, both in the ideal sense and in the sense of a distilled essence; as a generalized particular interest; etc.

Sometimes explicitly, sometimes by implication, in every way of thinking and talking about the public interest there is a place for the private interest, or interests, those of particular groups—economic, social, political, ethnic, geographic, religious. What we call the political process, including the governmental element, is a pragmatic way of identifying and integrating the large number and variety of particular interests, with the public interest as the result. Thus, at any particular moment in our public life the public interest is best thought of not as a single monolithic thing, but as the interests of a large number and almost infinite variety of goals and methods representing the hopes, the aspirations, and the activities of a large number of more or less discrete publics. Each public, or group, sets goals for itself. Each mobilizes resources, more or less rationally, to promote and, if possible, reach its goals. Each is drawn into the public arena, if reaching its goal or goals necessitates public approval or acquiescence. Each engages in the political struggle in elections and in Congress to the extent and in the way best calculated to maximize the achievement of its goals, stated or tacit. Our government, federal in form and representative republican in nature, provides the institutions and methods by and through which private groups and interests can obtain public sanction for their programs.

Recently the accelerated growth of science-based technology has had a profound effect on the process, particularly in extending to government and politics the specialization so characteristic of science-based technology itself. Each science with its corresponding technology regards itself as related to, but somehow different from, every other science, or branch of science, thus acquiring an identity of its own, with its self-regarding concerns, interests, hopes, and fears. In this way, too, they become recognized and represented in many of our basic institutions—in government, in the universities, in business and the economy. Of course, not every branch of science is recognized and represented in this way. Nor is pure science accorded the same recognition and representation as applied science and engineering. As a rule the more closely science-based technology can be identified with national security and defense the more quickly and the more completely it becomes accepted as a matter of national political and governmental concern. At the same time the dividing line between public and private science and technology, between research and development for defense and security purposes, and for nonsecurity or private purposes, becomes increasingly blurred and tends to disappear. When about half of industrial research and development is financed by the government; and when production facilities and working capital, particularly in the aeronautic, space, electronics, and shipbuilding industries, are owned more and more by the government, the private component of the public interest becomes increasingly difficult to identify and differentiate.

From the practical point of view, private entities, mostly corporations, are most interested in prompt action on their applications for permits, licenses, and other administrative actions. As a general statement this appears unexceptionable. It is in the application where difficulties arise, difficulties of such magnitude and variety as to throw doubt—almost—on the validity of the statement itself. In one way or another nearly everything said in the rest of the first part of this book illustrates the difficulties.

EXPERIENCE IN ASSIMILATING TECHNOLOGY INTO PUBLIC POLICY

In terms of environmental policy, technology is forcing a change from negative policy-making, or policy-making by default, to positive policy-making. From the beginning of the industrial revolution science-based technology has confronted government with two questions: Can government perform its functions better with new ways of doing them? And, how will government take account of the consequences, particularly environmental consequences, resulting from the use of technology on an ever-increasing scale? In supplying answers to these questions we will set out a body of experience acquired by the government in dealing with technological change.

In the first place, in the early years government was little concerned by the consequences of technology to the landscape, confining itself, in expanding and developing a continent, to providing inventors with a climate conducive to invention and guaranteeing them the right to enjoy the fruits of their creativity, and entrepreneurs a climate to exploit the country's natural resources, and settlers a chance to acquire land at little cost.

However, with respect to strictly governmental functions, such as national defense, the federal government was not long in adopting technological change: steam for sail, iron ships for wooden ships, in the case of the navy, in the mid-nineteenth century; later, with the development of the internal combustion engine, replacing the horse with motorized transport and artillery; still later, in the early twentieth century, experimenting with both heavier- and lighter-than-air craft, finally adopting the former; and in recent years, adding jet aircraft, rockets and missiles to the country's weapons systems. Atomic energy placed the government in a quandary: monopolize, or share with private industry? But not for long. After deciding to monopolize the atom for military purposes the government adopted a compromise permitting private industry to exploit it for peaceful purposes, particularly for generating electric power. Most recently space satellites, with the necessary launching systems, have been adopted, partly for peaceful, partly for military purposes. Together with rocket launchers and control capsules for space probes, space satellites can be and are communicated with and controlled electronically, thus meeting the necessary condition for success.

Research in the agricultural sciences was begun by the federal government over a century ago, and, since the 1880s, has been carried on cooperatively as well with the states. Likewise in medicine and health, the federal government in the 1950s began and has expanded rapidly investigations in mental health, cancer, heart disease, etc. Changing technologies in the fields of foods, drugs, and cosmetics have been reflected in the federal government's assumption of responsibility, in the name of the general welfare, for their purity, efficacy, and safety. Thus, over the past two centuries and particularly in the last fifty years the government's answer to the ever-quickening expansion of technology has been to absorb the new technologies and adapt them to the performance of traditional governmental functions.

But government's response to technology did not end there. At the same time government was answering our first question, adopting new technologies the better to perform its traditional functions, it has found it necessary, desirable, or expedient, to assume an almost totally new function—to regulate the products and the services made possible by the new technologies. Using a variety of old and some new methods the federal government now regulates the marketing of products of the food, drug, and cosmetics industries; tests for the safety of additives in the case of foods, for sanitary conditions in slaughter of animals for meat, for potency, effectiveness, and safety in the case of drugs, and for safety in the case of cosmetics. Also, products of industries derived from new technologies, advertising and broadcasting, for example, likewise come under federal regulation, the former, together with distributors, being prohibited from engaging in false and misleading advertising, the latter holding licenses of convenience and necessity subject to renewal and revocation. As with products, so it is with services: transportation, including transportation by rail, bus, or truck, pipeline, barge, or air, and communications, including communication by voice, wire, cable, or radio, all coming under federal regulatory jurisdiction. Not all services spawned by the revolution in communications are federally regulated: public relations, for example. So extensive has been the widening of federal regulation, however, that it may not be too sweeping an assertion to attribute to science-based technology the creation of a fourth branch of government—the independent regulatory agencies.

In answering the second question, however—how has government taken account of the consequences, particularly to the environment, resulting from the use of technology?—government has been slow in responding, negligent, and, until recently, generally unresponsive. Even now, with the National Environmental Policy Act on the books, it is by no means clear whether government has made policy (except formally); whether the study of ecology is more than a passing fad; whether the economy can and will accommodate itself to the steps needed to clean up the environment and reduce pollution; in short, whether interaction between environmentalists, government, and the other segments of society can bring about positive environmental policies which will be acceptable to all parts of the community. While there are numerous favorable signs that can be pointed to, on balance the future does not look very promising, and it is probably too early in the environmental revolution to appraise accurately the results. Changes in environmental protection or restoration standards are being made frequently.

2.
Science-Based Technology and Public Policy

A look at science-based technology and public policy in general discloses a number and a variety of methods, which Congress and the states have resorted to in attempting to keep abreast of the revolutionary effects of science and technology on life in the United States.

Looking at the narrower problem of technology and environmental policy in terms of patterns, procedures, and devices, both convenience and clarity of presentation might suggest listing and classifying the methods evolved over two hundred years.

No such attempt is made here. In fact, it is doubted whether either body of experience has received systematic consideration in these ways. In other words, so far as is known, no inventory has been attempted of the experiential data of the impact of science and technology on policy accumulated by the American people over their nearly two centuries as an independent country. Of course, it is generally known that the Constitution provides for a federal form of representative government; that Congress enacts laws; that they are entrusted to the chief executive and the bureaucracy to enforce; that private rights and disputed points of jurisdiction between the federal and state governments are decided by the courts. It is known, too, to most people that what amounts to an additional branch of government—the independent regulatory agencies—has been set up by Congress over the years. From this recital a rough categorization of methods emerges. But something more is needed. To benefit from the experience of history a listing, but not chronologically, of the patterns, procedures, and devices which have been adopted or have developed over the years may constitute a step toward systematic consideration of how to deal with these urgent public questions.

PATTERNS, PROCEDURES, AND DEVICES

Item 1. An early device or pattern adopted by the federal government was *government ownership and operation* of facilities, such as armories, naval yards and docks; research laboratories in agriculture, transportation, especially highways; and atomic energy, the latter operated sometimes directly, sometimes indirectly through a consortium of universities.

Item 2. Later another pattern was adopted—*cooperative federal-state programs,* as in agricultural research, financed by both federal and state funds, with states meeting federally-fixed standards. In essence, and due to our federal form of government, cooperative federal state programs have become, over the years, one of the characteristic patterns developed by Congress for meeting state and local needs. Present-day federal programs for air and water pollution abatement are based on this pattern. While there are advantages in the use of this pattern in some areas, such as agricultural extension and research, disadvantages are encountered, too. One problem is the representation of polluters on state regulatory boards. In 1971, of the fifty states nine had no boards (air and water pollution regulation handled by a full-time state agency); thirteen had combination air-water boards on which pollution sources were represented (industry, agriculture, county and city governments); sixteen had waterboards on which such pollution sources were represented; eighteen had air boards on which such pollution sources were represented. Five had water boards on which there was no such representation; five had air boards on which there was no such representation. On five combined air-water boards there was no representation of these major polluters.[1]

Item 3. In the 1930s the *government corporation,* wholly owned and financed by the federal government, began to be used,

---

[1]Some of the problems are adverted to later. See pages 49-50.

as for example the Tennessee Valley Authority, Commodity Credit Corporation, etc., or with mixed public/private ownership and operation—Communications Satellite Corporation (COMSAT).

In the 1940s Congress refused to adopt legislation for a Missouri Valley Authority (MVA) patterned after the Tennessee Valley Authority. Many competent observers, both here and abroad, regard the Tennessee Valley Authority as outstanding in upgrading the quality of the environment in the Tennessee valley and of life generally. Indeed, as an example of what Powell had envisaged a half century earlier for the great river valleys of the West; and as a showpiece of what a federal system of government can do, the TVA might well be considered a model for the Missouri River basin and for the country as a whole.

Item 4. Cooperation between what economists call *the public and private sectors* is a traditionally-sanctioned device used by the federal government to obtain the cooperation of private enterprise in furthering government policies. As already noted, the government contract was first used in 1792 to establish post roads and riders for carrying the mails. Since then it has grown to almost unrecognizable proportions. Today it is widely used, notably in the research, development, testing and production of naval ships, aircraft, rockets and missiles, with industry and the universities, both public and private, participating. Wartime performance in inventing and developing new weapons for the armed services under the Office of Scientific Research and Development (OSRD) suggests the desirability of examining the experience with this organization as of value in tackling the needs of environmental betterment. This is done below (pages 103-5).

Item 5. Fixing of deadlines, by statute, for performance in the private sector of business and industry is a modern device used by Congress in connection with attempts to alter the effects of technology on the environment. As a means of enforcing compliance with federally-determined standards Congress in 1970 passed the Clean Air Act. Full financing and vigorous enforcement of this law will have far-reaching effects on industry and on the country at large. Among other things this legislation tries to force the automotive industry to produce not later than 1976 automobile engines which will emit 90 percent less toxic exhaust fumes than at time of passage. By the same token, the failure to enact, before 1970, such legislation, federal and state, provides an example on a nationwide scale of policy-making by default.[2]

Item 6. Congressional attempts to fix standard operating procedures (SOP) in the executive branch illustrates both a newly-established pattern and a new procedural device. In the 1969 National Environmental Policy Act Congress legisled procedures requiring each department and agency in the executive branch to submit to a newly-established Environmental Quality Council comment in the form of environmental impact statements on all legislative proposals which might affect the quality of the environment. (This is a variation, done by Congress, of Budget Bureau directives of the 1930s: no legislative proposals to be originated in the executive branch or submitted to Congress without clearance by the Bureau of the Budget.) In retrospect, the 1969 legislation may turn out to be the initial attempt at technological assessment by the federal government (see below, pages 52-57).

Item 7. Over a long period of time the procedural device of *administrative hearings* according to a pattern fixed by statute has evolved as a method more and more frequently resorted to in our attempts to reconcile private rights with the public interest and make environmental policy more responsive to technological change. In general, the regulatory agencies, which are largely independent of the executive branch, and certain administrations (Food and Drug, for example) and bureaus (Animal Industry, for inspection of meat in interstate commerce), are required by Congress to follow this procedure. However, it is not required for the Department of Defense, where cost-plus and/or negotiated contracts are the rule. Nor does the Federal Highway Administration, in administering Highway Trust funds; the Office of Education, in administering federal aid-to-education funds; or the Social Security Administration, in administering Aid to Dependent Children

---

[2]Wayne T. Sproull, *Air Pollution and Its Control,* Second Edition, Revised and Enlarged, New York: Exposition Press, 1972, especially Chapter 9, "Legal and Economic Aspects."

(ADC) funds, have to hold public hearings as a prerequisite for disbursing these funds (see pages 33, 50-51, 55-56).

Item 8. The use of *scientific findings and technological data* as a basis for legislation, for administrative action, or for influencing public opinion is a device increasingly employed, sometimes alone, more frequently with other devices. Legislative committees regularly develop such reports as part of their fact-finding function incident to preparation of legislation. Scientific and technical reports by administrative agencies are used differently by different groups. The Report on Cigarette Smoking and Health of the Public Health Service's Surgeon General is an example. Here some bureaus of the federal government allied themselves with the tobacco manufacturers and the broadcasting industry in their delaying action, while others, with various allies, gave vigorous support to a regulatory and, ultimately, a prohibitory, policy.

Item 9. Buying scientific and engineering expertise either directly, as when government employs scientists and engineers, or indirectly, when, through contracts and grants, the government acquires their technical knowledge. Even before the dawn of the atomic and space ages, in 1945 and 1957 respectively, the government, during World War II, had expanded greatly its use of this method of acquiring scientific knowledge and management skills. OSRD was the government agency in charge. Its phenomenal success in innovating new weapons to meet specific needs of the armed services suggests once more the desirability of examining its philosophy and organization in connection with the environmental crisis (see pages 102-3 and following). Today, the largest users are, probably, the Atomic Energy Commission (AEC), the National Aeronautics and Space Agency (NASA), the National Institutes of Health (NIH), and the departments of Defense and of Transportation (DOT). Less widely used are contracts by one government agency to get the expertise of another, e.g., the Food and Drug Administration (FDA) in the Department of Health, Education, and Welfare (HEW) contracting with the National Academy of Sciences (NAS) to test the effectiveness of drugs.

Advice from scientists and engineers is solicited far more widely by the government's executive than by its legislative branch. There are many more scientists and engineers employed by the agencies and departments directly. These number in the thousands, serve on hundreds of committees, and advise the Department of Defense and other departments concerned with scientific, technical, and other questions. In addition, through the National Academy of Engineering and the National Research Council, the National Academy of Sciences, a semipublic institution, oversees the work of some seven thousand engineers and scientists who give part of their time on about five hundred committees. Not included are those in research institutes ("think tanks") which have been set up to advise the air force, the army, the navy, and the Department of Defense. All in all between fifteen thousand and twenty thousand individual scientists and engineers are involved in this scientific advisory system. By comparison the legislative branch—Congress —has the advice of a few dozen scientists and engineers, if that many, in its committee staffs and in the Library of Congress Legislative Reference Service.

The gravity of the situation which confronts government agencies in administering and enforcing laws regulating the interface between science-based technology and the environment is well illustrated in the pharmaceutical industry's relations with the Food and Drug Administration. Ten years after Congress amended the Food and Drug laws in 1962 to take account of the rapid increase in the number and variety of synthetic drugs FDA was faced by what its counsel asserted was "the biggest challenge it has ever had to its ability to assure the safety and efficacy of medicines."

Unlike the FDA, which deals with food, drugs, and cosmetics used by people, the Atomic Energy Commission deals with the uranium mining and processing industries and licenses and regulates atomic reactors.

In both cases, however, the regulatory function includes the developing and fixing of safety standards—FDA so that drugs, etc., will be safe for humans; AEC so that nuclear reactors and other uses of radioactive materials can be operated without danger to both humans and the environment. Bothering both the FDA and the AEC is the fact that both the pharmaceutical industry and the nuclear power industry are new industries, in the sense that safety tolerances are not yet known with precision. And the chemists working for FDA and the National Academy of Science, which works for FDA on contract, and the technologists working for the AEC do not agree with medical doctors engaged by pharmaceutical companies and with other technologists as to what constitutes safety. Meanwhile, the highly competitive pharmaceutical industry markets its products pending FDA determination of purity, efficacy, and safety, and the unparalleled increase in the demand for electrical energy places the AEC under heavy pressure to license construction and operation of new generating plants while questions of protection from radiation and precautions against accidents are still being debated. Such are some of the extremely complex scientific and technical problems uncovered when government undertakes to protect the public interest while refraining from undue interference with growth and progress.

Item 10. In a sense the whole movement of *governmental reform,* particularly of the legislative and executive branches, may be regarded as an effort to strengthen the federal government the better to enable it to keep abreast of technological change. By adding new committees to its structure Congress acknowledges technological innovations, tries to inform itself about them, and enacts new statutes for dealing with them. Thus, the 1946 Atomic Energy Act provides for a joint Senate-House committee on atomic energy, which was followed by separate Senate and House committees on science and astronautics which, in turn, played a prominent part in shaping policy toward space when the Russian orbiting of Sputnik in 1957 seemed to require a new organization within the governmental structure. The consequence was the National Aeronautics and Space Administration. Congress also has reformed and strengthened its own staff agencies—the staff services of its committees, the Library of Congress Legislative Reference Service, and the General Accounting Office.

Positively, by delegating authority to the chief executive to reorganize, or by acquiescence, Congress has strengthened the executive office of the president, with the establishment of the Office of Science and Technology (OST), the office of the president's science advisor, and the president's Science Advisory Committee, with numerous panels, and the Council on Environmental Quality (CEQ).

As part of the movement of governmental reform the legislative process itself has been altered, although not radically. The Joint Congressional Committee on Atomic Energy, already noted, is a structural and procedural innovation which is unusual. The only other joint congressional committee is that on the economic report, set up by the Employment Act of 1946, a tacit, if not overt, acknowledgement of the need of Congress to update itself in trying to deal with the economic problems of a technological era. However, it is noteworthy that not even today, when atomic energy and the space sciences applied to military purposes have revolutionized strategic concepts, has Congress moved to modernize itself by establishing a joint committee on national security (or the national defense or the armed services).

In still another sense Congress has been slow in adapting itself to the changes brought about by revolutionary technology. Procedurally, Defense Department budget estimates running into the scores of billions are still examined as if they were of pre-World War II magnitude. In recent years Congress has taken only two short and inadequate steps to modernize itself (1946 and 1970). Moreover, it has been unable to deal with the tendency of the regulatory agencies to reflect the desires of the industries they are supposed to regulate to escape regulation, if possible, or, failing that, to come under the influence of these industries. For example, in helping the railroad industry get rid of passenger service the Interstate Commerce Commission (ICC) was overzealous. The Federal Communications Commission's (FCC) policing of the publicly-owned airwaves is not too diligent. The Federal Power Commission (FPC) and the Federal Aviation Administration (FAA) have not been overly imaginative in regulating power and gas and aviation industries. The same is true of the Food and Drug Administration, as a pioneer study of the influence of the pharmaceutical industry on the FDA in developing new drugs shows.[3]

---

[3]Morton Mintz and Tim O'Brien, "The Guinea Pigs," *The Washington Post,* October 24, 1971. The guinea pigs are human beings: ". . . the drug companies, with the collaboration and collusion of the FDA, are really doing their experimentation with new drugs on humans while simultaneously carrying on limited and minimal animal studies on the side as a façade."

Part of the explanation lies in the failure to differentiate between promotion and regulation, a blurring of the problem of technological assessment most bafflingly exemplified perhaps in the Atomic Energy Commission.

Another part of the explanation lies in the numerous puzzling problems accompanying any governmental attempt to regulate new industries which are doing their best to escape effective regulation. In the case of cable TV (CATV) both intraindustry and federal-state rivalries complicate the regulatory problem.

Still, it is most likely not a failure to differentiate between promotion and regulation. Almost always congressmen know what they are doing when new technologies seeming to require consideration are brought to the attention of Congress or when environmental degradation becomes so palpable that it cannot be ignored by Congress. In these circumstances the legislation issuing from Congress is rarely based on a rational choice between competing alternatives. More likely, it is the result of bargaining and compromise under bureaucratic and industry pressures, both of which are characteristics of the legislative process in the United States.

Item 11. Bottlenecks and Gatekeepers. In the pre-legislative and legislative phases of national policy-making *personalities* often loom larger in the outcome than other more obvious factors. Of the people most directly involved in funding environmental programs in the first session of the 92d Congress (1971), Jamie L. Whitten (D., Miss.) is no doubt the key man. Next to him comes George H. Mahon (D., Tex.), chairman of the House Appropriations Committee. On February 10, 1971, when the 92d Congress was organizing, he named Mr. Whitten chairman of the Appropriations Committee's agriculture subcommitee and had placed under its jurisdiction the appropriations for the Environmental Protection Agency, the Council on Environmental Quality, the Food and Drug Administration and the Federal Trade Commission. In the policy-making process, therefore, these two strategically-placed congressional leaders outrank William D. Ruckelshaus, administrator of the Environmental Protection Agency, Russell E. Train, chairman of the Council on Environmental Quality and the president's chief environmental aide, and even the president himself.

On the same day that the press carried news of the expanded jurisdiction of Mr. Whitten's subcommittee Mr. Train contributed a piece, "Toward a Better Environment," summarizing in promising language the president's recommendations to Congress.[4] To those who follow the in-fighting surrounding environmental programs the news of Mr. Whitten's added authority came as a discouraging development. Specifically, the League of Conservation Voters, which follows votes in Congress on key issues, gives Mr. Whitten a strongly negative rating. He is known to environmentalists, too, as the author of a book, *That We May Live,* published in 1966. It is based on a study of the effects, uses, control, and research of agricultural pesticides made at Whitten's request in 1964. The book has been called "unscientific," "uncritical assimilation of material which marred the original report," and "clearly an answer to Rachel Carson."[5]

The treatment of the president's requests for funding is shaped in the House appropriations subcommittee. These requests include money for EPA, CEQ, FDA, and the FTC, as well as for the proposed new programs for controlling dumping in oceans, the Great Lakes, and estuaries; for the development by the states of land-use plans; for advanced planning of power plant sites, strip mining regulation and the use of taxes to discourage development of wetlands and encourage preservation and rehabilitation of historic buildings. Mr. Whitten presides over this committee.

Item 12. Interstate compacts, comprising simultaneously a pattern for action, a set of procedures, and a potentially valuable device for environmental improvement, are provided for by the Constitution for promoting cooperation among the states and the federal government. Over the years resort to this complex of procedures has not been great. Possibly its longest use has been in the allocation of the waters of the Colorado River among the various riparian states. Here, the value of the interstate compact as a means of preventing environmental degradation and of ar-

---

[4]*The New York Times,* February 11, 1971.
[5]Frank Graham, Jr., *Since Silent Spring,* Boston: Houghton Mifflin Co., 1970, p. 181.

resting improper land-use practices has been minimal. While the vision of John Wesley Powell, explorer of the Colorado River Basin, for development of the Southwest has been realized, it has far outstripped the water resources available for development, with the result that the interstate compact device has been inadequate as an effective means of environmental protection and improvement. On the contrary, rivalry among the several states, competition for water, burgeoning populations and a growth philosophy have resulted in almost continuous litigation since the first compact was negotiated in the 1920s.

Where these conditions have not prevailed, at least not to the same degree, the interstate compact has proved more satisfactory. With the approval of Congress, the states of New York, Pennsylvania, New Jersey, and Delaware joined in the Delaware River Basin Compact. Under a four-state commission, on which the federal government is represented, these four states have an opportunity to achieve substantial benefits to the environment. The opportunities have been underlined in 1971 in a study of the next ten years of the Delaware River Basin Compact in the light of the National Environmental Policy Act made for the Commission by the University of Pennsylvania's Institute for Environmental Studies.

But to expect from interstate compacts comprehensive treatment of environmental problems is like asking a boy to do a man's job. Even under optimum conditions they do not work well. Environmental quality is too many-sided, too complex, and affects too many segments of the community. However, when states supplement their compact by imaginative and far-sighted legislation, the chances for success are increased. Delaware has done this. Under its urging New Jersey is considering enabling legislation based on Delaware's coastal zone control legislation. Maryland, too, is showing interest in this promising tool of environmentalism.

The Delaware Story. The enactment by Delaware in 1971 of legislation barring further heavy industrial development in its coastal zone may prove to be a model for both federal and state governments in dealing with the environmental crisis.

Certain industries—oil refineries, basic steel manufacturing plants, basic cellulose pulp paper mills, and chemical plants such as petrochemical complexes—are banned in the future, as well as offshore gas, liquid, or solid bulk product transfer facilities. The law "seeks to prohibit entirely the construction of new heavy industry" in Delaware's coastal areas.

Other industrial developments, such as auto assembly plants, garment manufacturers and similar "clean" facilities, are permitted, but only after the issuance of a permit by the State Planner under guidelines adopted by a Coastal Zone Industrial Control Board, provision for which is made by the newly-enacted law. The Board also hears appeals from decisions of the State Planner in administering regulations implementing the legislation.

Although the full story of the enactment of the law remains to be told, it is known that Governor Russell W. Peterson, a Ph.D. in chemistry, led the fight for enactment.[6] He set up the task force which recommended the enactment of this unprecedented legislation; guided it through the state senate and house of the General Assembly; rebutted the arguments of federal officials opposing the law; and signed the law. He has also appointed to the Control Board a representative of the environmentalist point of view.

In its approach to the future of Delaware's coastal areas the law is positive as well as negative. It is stated clearly that the use of beaches, marshes and open space for recreational purposes is to get first consideration over promises of financial gain contained in proposals for further economic growth as understood heretofore.

In passing it may be noted that of all the practices, procedures, and devices enumerated and discussed herein, the interstate compact under enlightened leadership shows more promise than any other in approximating our ideal of rational development of resources, particularly water, through regional-river basin development. The Tennessee Valley Authority shows what can be done even under our rigid federal system. Not dissimilarly, the Delaware

---

[6]"He Slammed the Door in Industry's Face," interview with Governor Russell W. Peterson, *National Wildlife*, December 1971-January 1972, pp. 50-51.

River Basin Commission and an even more recent one, the Susquehanna River Basin Commission, established in 1971 by New York, Pennsylvania, Maryland, and the United States, could be the forerunners of others constituted under the interstate compact clause of the federal Constitution.[7]

EMERGING NEW PRACTICES

Resort to the courts, that is, litigation, is a practice which, until recently, has not been used to the same extent as the patterns, procedures, and devices set out above. Historically this is so, not only because environmental policy as a public problem and issue has surfaced only within the last decade, but also because jurisprudence has been slow in making a place for the concept and for the term in American constitutional law and practice. All this is changing. Some time in the near future, the time will have arrived when litigation and administrative hearings resulting in litigation will be recognized and generally accepted. That time is not yet. While a few judges have been receptive to actions brought in their courts on behalf of classes and of the public as a whole, far more have been unreceptive or have not yet been called upon to deal with the matter.

To illustrate, in a case before the Federal Court of Appeals for the District of Columbia in 1971 the role of the Federal Power Commission relative to the environmental impact statement requirements of the National Environmental Policy Act of 1969 was clarified. At what stage in the FPC's licensing procedure for a transmission line should the statement be introduced? Before or after the Commission's trial examiner conducts hearings? Does the law require the FPC to prepare its own statement or is it enough to use and introduce one prepared by another agency, in this case

---

[7]It was reported on January 16, 1973, that Governor Nelson D. Rockefeller (R) of New York had hired former governor Russell Peterson (R), who was defeated for reelection in November 1972, to head a new state commission to study "the role of the modern state in our changing Federal system."

the New York State Power Authority? Who can raise such questions? What effect will the court's decision have on the FPC's procedures? The presiding judge in the case, which involved a proposed transmission line from a pumped storage facility along Schoharie Creek in upstate New York to the town of Leeds, thirty miles south of Albany, decided that the FPC had followed an incorrect procedure, both as to whose impact statement should be prepared and introduced into the record and when it should be done. It was not correct, the judge said, for the FPC to circulate someone else's statement; its staff should prepare its own. Moreover, it was not enough to introduce the statement after the trial examiner had conducted hearings; it should be done before, so that petitioners in the case, the Green County Planning Board, the Town of Durham, and others could examine it before hearings were held, could call witnesses and cross-examine them during the hearings. In his decision the judge is reported to have complained that "the Federal Power Commission has abdicated a significant part of its responsibility by substituting the statement of (the State Power Authority) for its own."[8] The judge rejected other requests of the petitioners, that the whole project be blocked, and that court costs be paid by the federal or state power authorities.

This federal district court decision was appealed by the Federal Power Commission but was upheld by the Circuit Court of Appeals. And the Supreme Court did not agree to review the Court of Appeals decision. In late 1972 the FPC announced it would henceforth require its own staff to prepare environmental impact statements in transmission-line construction applications.

Thus the Federal Power Commission is being compelled by the courts to provide a larger place in its procedures for the environmental impact statements required by the Environmental Policy Act. Similarly, the judiciary, when asked by interested citizens, is slowly clarifying what environmental impact statements require from other agencies. The lengthening list includes the Atomic Energy Commission, the secretary of agriculture and the Forest

---

[8]*The New York Times,* January 31, 1972.

Service, the secretary of the interior, as well as the Federal Power Commission.[9]

At the same time the federal Department of Justice claims to be making new law when it asks the courts to find a mining company guilty under the Refuse Act of 1899 of committing a public nuisance when it dumps mine tailings into Lake Superior.[10]

One of the traditional roles of law in America is as an agency for social change. This being so, it could have been predicted that the courts would be called upon to play an expanding part as society's cutting edge in the confrontation between science-based technology and environmental policy. At least for the immediate future it appears that the courts will be called upon more and more to perform this function. The National Environmental Policy Act heralds this increase in the part to be played by the judiciary. And federal legislation regulating power plant siting, to supplement state legislation regulating such siting and location of transmission lines as well, as in New York State, pends before Congress. These are parts of a problem of a higher order of magnitude, a national energy policy, which is discussed in chapter 4. Thus, in addition to the milestone Environmental Policy Act and the court decisions which are based upon it, a continuing effort to write and implement environmental policy is going forward.

So far as use of the courts in making environmental policy is concerned there are at least two viewpoints. One holds that law is *the* agency to bring technology under control; that without it technology becomes exploitation rather than progress; and that man's intelligence, when he uses it through law, can tell him when it is time to stop.[11] The other viewpoint holds that litigious methods are time-consuming, costly, and spotty in effect.[12] Without taking sides, it seems clear at least that science-based technology, willy-nilly, confronts mankind with painful decisions affecting the hu-

---

[9]*The Wall Street Journal,* February 2 and March 6, 1972; *American Forests,* vol. 77, nos. 6 and 7 (July and August 1971).

[10]*The Washington Post,* May 7, 1972.

[11]David Loth and Morris L. Ernst, *The Taming of Technology,* New York: Simon & Schuster, 1972.

[12]Personal communication from E. W. Morehouse.

man environment in the making of which law is available for help. If and when Congress or the states or the executive branch of government acts under the Constitution or the laws and such acts are held by some to be detrimental to the environment, the courts are certain to be drawn into the process for help in making decisions.

POLITICS

As one category in an enumeration of means providing in some detail how government has assimilated into policy the technology which has affected our environment so profoundly, politics appears appropriately toward the end of the list. Not only does politics provide the setting within which all the other patterns, procedures, and devices operate; politics also is the method *par excellence* of adopting, adapting, and sometimes, but not frequently, reversing a course of social action.

As already noted (items 10 and 11), all legislation and especially legislation having to do with the environment, runs anything but a linear course. The conventional wisdom about making a law is belied at almost every stage. Bureaucratic and industry pressures are directed against legislators. The Executive Office of the President, the president's staff arm, is a force, positive, negative, or neutral. In Congress there are bottlenecks and gatekeepers, leaders and groups who stand at the strategic points in the process and make the crucial decisions. Numerous segments of the public, scientists, writers, the news media, the many publics interested for a variety of reasons in the outcome, ply their various crafts and professions attempting to influence the outcome. A major force is propaganda, the management of collective attitudes by the manipulation of significant symbols. Partisanship is often of importance, not so much the electoral partisanship of the political parties, although on occasion this is a factor, as the partisanship of individuals, leaders or would-be leaders ambitious for political office. Back of this, and underlying it all, is the general public, the undifferentiated mass of people marginally interested in the issue at best, mostly apathetic or uninterested and feeling uninvolved at the worst, unless and until their economic interests appear threatened. And the relative position of the issue on the public agenda must not be overlooked. The pending bill is competing with all other pending laws, all major domestic and foreign events, all local, national, and foreign policy issues, all attempts of the chief executive to manipulate public opinion to his and/or his party's advantage. Much of the bargaining, the maneuvering, the pressures, compromises and decisions takes place underneath the surface, out of the glare of the media.

In recent years environmental policy has surfaced as a political issue. In each of those years (1969-72) President Nixon sent special messages to Congress asking the enactment of numerous pieces of legislation, all designed to stop environmental degradation or, at least, to placate importunate pressure groups. Although as an issue environmental policy appears not to have been of consequence in the 1972 presidential campaign, it did figure in congressional and state and local contests. Of fifty-seven candidates for the House, the Senate, and governorships, of whom thirty-five were Democrats and twenty-two were Republicans, who were endorsed by an environmentalist group, the League of Conservation Voters, forty-three won. Together with pro-environmentalist votes in referenda in California, Washington, New York, and Florida, these successes prompted *The Washington Post* to say (November 11, 1972): "It is clear now that concern for the environment is not the fad or passing whim many believed it was."

Specifically as regards legislation, changes in the federal pesticide law, pending in Congress in 1971 and 1972, were the subject of intense buffeting from numerous sources and of equally active bargaining in a number of different forums. The major issue was a strong or a weak law, that is, continuing and expanding federal regulatory authority over pesticides or lessening and weakening the existing authority. Even while Congress was pondering this issue, numerous federal departments and agencies were tied up in an attempt to enforce the legislation already regulating the use of DDT and other persistent hydrocarbons already on the statute books. These included the federal courts in the District of Columbia, the secretary of agriculture, and the administrator of the Environmental Protection Agency, and further among several bureaucracies those in Agriculture and in EPA. Agricultural chemical manufacturers were making strenuous efforts in the courts, administrative hearings, among legislators, public opinion leaders, and the news media to block administration efforts to cancel DDT for all but essential uses and, simultaneously, influence key congressmen and public opinion in their deliberations on pending pesticide legislation.

On October 7, 1971, Dr. Norman E. Borlaug, a plant geneticist and Nobel Prize winner,[13] appeared at a DDT cancellation hearing in Washington at the invitation of the United States Department of Agriculture and, shortly thereafter, at a news conference arranged by the Montrose Chemical Corporation, world's largest manufacturers of DDT. On the crucial subject of the effect of DDT on wildlife, Dr. Borlaug was disqualified as an expert witness by the trial examiner on the ground that he was not an expert on that subject. According to Dr. Borlaug's own testimony, he had last studied pesticidies in 1944 at a time when he was employed by a pesticide manufacturer. From this hearing Dr. Borlaug went to the press conference arranged by the Montrose Chemical Corporation where he is reported to have made "unsupportable assertios about DDT and other pesticidies, wildlife, cancer, insect control, hysterical environmentalists and other subjects he was not competent to discuss in the hearing. His statements, of course, have been exploited for publicity purposes by the pesticide industry."[14]

Extensive comment on Dr. Borlaug's performance at that stage in the hearings and in the congressional consideration of amendments to pesticide legislation is hardly necessary. Suffice it to say that as with all federal dealings with the regulation of industry, it is particularly important in the shaping of wise and prudent legis-

---

[13]In 1971 Dr. Borlaug was awarded the Nobel Prize for Peace.
[14]Charles F. Wurster and Robert van den Bosch, "The Incredible Borlaug Blunder," *Audubon,* January 1972, vol. 74, no. 1, pp. 118-19. Dr. Wurster is associate professor of environmental sciences at the State University of New York at Stony Brook. Dr. van den Bosch is professor of entomology at the University of California at Berkeley.

lation concerning the environment to determine who is using what allegedly expert witness for what purposes.

TREATIES AND EXECUTIVE AGREEMENTS

Treaties and executive agreements constitute still another means for bringing governmental authority to bear on the effects of technology on the environment. Before the application of atomic energy to military purposes (1945) the record shows few examples of this use of the treaty-making power. Twenty-five years earlier the use by the federal government of power to protect birds summering in Canada but wintering in the United States was successfully asserted. By 1945 the question had become, not only the protection of birds, but also the protection of the biosphere, including man himself, from radioactive fallout. There was no question here, as there was in the Migratory Bird Treaty matter, as to who had the authority, the federal government or the states. At least until the mid-fifties, no state, and few, if any, individuals or corporations disputed the exclusive power of the federal government to enter into treaties reducing the threat of radioactive fallout. Using its power the federal government has entered into treaties banning nuclear testing in the atmosphere, on the surface, and in the oceans, but not underground, and proclaiming that outer space should be reserved for peaceful purposes only. Using this same power the United States agreed with the U.S.S.R. in 1972 to limit the permitted number of antiballistic missile sites and the numbers of permitted atomic weapons.

Also, in 1972, a United Nations-sponsored Conference on the Human Environment convened in Stockholm. There, 113 countries reached agreement in two weeks on 109 recommendations for international cooperation on environmental problems. So far as the United States is concerned, it is the federal government which, after the United Nations has considered the recommendations, will decide on which if any of them will be accepted.

Bilateral cooperation between the U.S. and the U.S.S.R. in numerous fields, including environmental matters, was furthered by the two countries at the Moscow conference in 1972. For the United States the device of the executive agreement, not depending on Congress for its implementation (except for appropriations), is being used to put into effect the agreements in these fields.[15]

In the governmental process, considered broadly, forces generated internally and externally act on and react to public opinion. Here occurs the ultimate test of the methods used by government to respond to the environmental crisis and for focusing science-based technology on extracting the malignancy from the crisis. We noted above that in this process the scientific advisory system should play a crucial, if not decisive, role. Embracing much of the scientific establishment but a much smaller part of the scientific community, the scientific advisory system is supposed to and in fact does advise the decision-makers with the best available scientific advice on specific technical questions as well as on broad public issues which are not wholly technical.

How well does it do its job? Of the many answers which might be supplied, one comes from a scientist who has worked within the system. Using as a test the decision-makers' acceptance of the advice tendered, he reports that on many specific questions the advice is accepted but that on broad general matters in which scientific questions are not the only ones present the system has not been a success. Such a question as "How does method A for water desalination compare in energy requirements with method B?" illustrates the first category of questions. For the broader questions he turns to such matters as the supersonic transport (SST), cyclamates, the safety of commercial nuclear power plants, the safety of underground nuclear tests, pesticides, and the use of herbicides in Vietnam. On such questions the recommendations of scientists brought to Washington to advise the administration are usually not accepted. They are, in fact, usually ignored, if there are strong pressures from special groups opposing the new policy, if the administration would be exposed to electoral or congressional difficulties if the advice were to be adopted, or if the advice is contrary to already existing administration policy.[16]

---

[15]Barbara Ward and Rene Dubos, *Only One Earth,* New York: W. W. Norton, 1972; *Audubon,* September 1972, vol. 74, no. 5, pp. 116-22.

[16]Martin L. Perl, "The Scientific Advisory System: Some Observations," *Science,* vol. 173, no. 24, September 1971, pp. 1211-15.

Of course the scientific advisory system as we know it today is the product of the last twenty-five years or so, the post-World War II period. Its failure or at best its limited success cannot fairly be held responsible for the present condition of the environment. However, considering the accelerated pace of scientific development in recent years the large gap between technically sound advice and political acceptability of that advice can hardly escape notice.

The discussion of patterns, procedures, and devices just concluded is limited to only the most obvious ways that government in the United States has reacted to environmental change brought about by successive technological revolutions. Nor is it complete. To mention only one radical change in governmental and business practices—computerization—is to open a wide range of problems with which both government and the economy are confronted, as a result of the post-World War II research and development of computers. However, no more is done here than to allude to it, since direct effects on the environment are small. Indirectly, of course, the computer industry, like air-conditioning and electric heating, are parts of the larger problem of the much greater demand for electrical energy with all that implies for environmental deterioration and improvement.[17]

---

[17]The vast and intriguing potentialities of computerization in the public sector, education, as distinct from its current application to the private sector—entertainment, banking, etc.—is well brought out by John G. Kemeny, in *Man and the Computer* (New York: Charles Scribner's Sons, 1972), especially his proposal on education, pp. 136-38.

3.

The Need to Sensitize and Inform America About Technology

In view of the existing state of the environment and the obvious failure of government to take adequate steps to stop environmental degradation, the great need is to sensitize and inform society and to sharpen its understanding of both the harmful and the beneficial in science-based technology.

OBLIGATION TO RECOGNIZE THE NEED

It is incumbent on the American people to rethink their attitude toward science and resources. The change needed should be revolutionary in scope, as well as in content, and it cannot be evaded with impunity. The civil rights movement and an unpopular war in Southeast Asia have sparked revolutionary thoughts, particularly among the young, even including resort to violence to correct alleged wrongs inherited from their forefathers. The thoughts evoked by the war in Indochina are now a decade old. Those evoked by discrimination against ethnic and racial minorities are the culmination of the frustrations of a century and more. But much farther back in origin and even deeper-seated are the attitudes, practices, and beliefs which must be changed if we are to adapt ourselves to the exigencies of an area extracontinental in territorial scope depleted by the ravages of a galloping technology. He spoke the truth who first noted that the real revolutionaries of our times are the scientists. The rest of us, then, are in good company if we insist on the necessity for equally revolutionary changes in thought and attitude toward our heritage of natural resources

41

and in our use of them. And in our preoccupation with making a living we must be cognizant of the malleability of technology to humanistic goals.

Much of the "in-fighting" over environmentalism (there is little dispute over whether there is a crisis) stems from a failure to examine the "inarticulate major premise" of our time (to borrow Justice Holmes's phrase), namely, attitudes as usual. Paraphrasing the phrase "business as usual" as applied to earlier crises, it is assumed, as it was in the prewar days of 1917-18, and, again in 1940-41, that crises in human affairs are nonrecurrent; that change, while inevitable, proceeds at a leisurely pace; and that the traditional beliefs about economics, politics, and pleasure need not be examined, let alone changed.

On the contrary, the many-sided crisis of our time, of which environmental degradation is the symptom, is of a kind and order of magnitude differing from previous crises.

With respect to the *kind* of crisis, from observation of the environmental degradation around us, and from reflection on its causes, one can readily conclude that resources—land, water, and air—have not only been used; they have been *mis-used*. Our observer can also conclude that earlier priorities, emphasizing amount of growth rather than quality, are in need of change. Also, for too long, subordination of the public to private interests has been the consequence of placing undue emphasis on individual rights at the expense of human rights. And preoccupation with growth accompanied by an obsession with the "can do/ should do" philosophy overlooks completely the inevitable antisocial consequences.

With respect to the *magnitude* of the crisis, our observer may well consider the size of the job of changing public attitudes toward resources; of changing traditional practices in agriculture, in mining, and in manufacturing; of adapting the outlook of government bureaus to the needs of the environment; and of maximizing the hidden potentialities of our federal form of government.

By referring to agriculture-related waste both the kind and the magnitude of the crisis can be dramatically illustrated. Many Americans are no doubt aware of industry and manufacturing as the source of much waste. Few, however, think of agriculture in this connection. Yet it is reliably reported that agricultural activities generate annually 2.5 billion tons of solid waste, more than half of all solid wastes from all sources. This is nearly ten times as much solid waste as is generated by all our cities, towns, suburbs and communities combined. And specifically in forestry, disposal of logging debris by prescribed burning produces more particulate air pollution than all our motor vehicles, industries, power plants, space heating and burning dumps combined.

In this connection consider how the new practice of finishing beef cattle in feed lots before slaughter has increased the pollution of streams in the cattle country. As long as organic fertilizer (barnyard manure) was returned to the soil of the individual farmer's ranch or feed lot, the practice was tolerable. In fact, such was considered good farming practice. Now, when thousands of cattle are collected in feed lots for finishing by grain-feeding accelerated by injections of rapid growth chemicals; and when the volume of organic refuse is multiplied a thousandfold and must be flushed into streams, feed lots become in effect a public nuisance. From the same authorities noted above we learn that some of the feed lots supplying us with steaks and hamburgers have sewage problems amounting to the equivalent of one million people living on 320 acres, a population density twenty-six times that of Calcutta.

Together with stream sedimentation caused by building construction (silt in runoff from new housing developments), feed lot pollution is expected by the President's Council on Environmental Quality to add $9.6 billion to the estimated $277 billion cost of cleaning up the environment over the next decade.[1]

In dealing with agriculture-related waste government has shown little initiative. At the moment what concern there is does not seem to be concentrated on action programs, either in the states or the federal government, but rather on research. A joint U.S. Department of Agriculture-Office of Science and Technology

---

[1]Gary Soucie, "How You Gonna Keep It Down on the Farm?" *Audubon,* vol. 75, no. 4, September 1972, pp. 112 ff.; Executive Office of the President, Council on Environmental Quality, Third Annual Report.

report in 1969 recommended a five-year program of research and development and action. But USDA expenditures in 1972 were only $3.03 million for R&D in animal waste management and $8.02 for action, out of $3.76 recommended for R&D and $178.43 for action. The year before, 1971, only $2.25 million was spent out of the $64 million appropriated of the $80 million authorized by Congress for agricultural waste projects under the Solid Waste Disposal Act.

Turning to industry and government, the picture is not encouraging. The 1967 Census of Manufactures (most recent available) discloses little which would support the belief that either government or industry is aware of the extent of environmental degradation and of the movement to stop it and reverse the trend. No data on either volume or value of pollution abating and control machinery are to be found in the most recent Census of Manufactures. One searches in vain in the industrial division of the Census Bureau's Standard Industrial Classification for any data on waste recycling and other environmental improvement systems. In fact, in Industrial Division (D) of the Standard Classification, pollution control equipment, waste disposal equipment, and waste recycling machinery do not appear as such in statistical tables for establishments by industry groups and industries.

In Volume I of the Census of Manufactures there is a chapter (7) on water use in manufacturing with a table (6) on water intake, water treated prior to use and number of establishments treating water prior to use by type of treatment for 1968; a table (7) on gross water used and number of establishments recirculating and treating water by type of treatment for 1968; and a table (8) on water discharged, water treated prior to discharge and number of establishments treating water prior to discharge by type of treatment for 1968.

Thus, these tables suggest how much more pollution would have been caused by manufacturing industry if water used had not been treated prior to discharge and by how many establishments and in which industries. Such data will no doubt become more plentiful as enforcement of old and new water-treatment legislation takes hold.

But the meagerness of water treatment data and the absence of any data on a pollution-abatement machinery industry seem to reflect the failure of industry and of government thus far to make a place for an environment-improvement-machinery industry in their current programs.

Nor does the record of industry in this connection provide grounds for great optimism. Some measure of the extent and gravity of industry-caused pollution, not of the physical environment but of people, may be gathered from an estimate of savings which could be realized from abatement of this pollution. Examining air pollution and human health in quantitative terms and estimating the savings which would result from a 50 percent abatement of certain kinds of air pollution, researchers think that a conservative estimate is $2,363,000,000 annually. Certain diseases only and certain sources of pollution only are considered. No attempt has been made, for example, to estimate the cost of relatively low levels of carbon monoxide in the air in terms of morbidity and mortality.[2]

Against such background the remark of a reporter for *Science* magazine assumes some substance. Glancing back over the post-World War II period, he says: ". . . (the post-war) record of government in guarding the public against the negative effects of technology has not been inspiring."[3]

SENSITIZING SOCIETY TO EFFECTS OF TECHNOLOGY

Assuming our ability to design an instrumentality for sensing and measuring the effects of technology on society, what would its principal specifications be? Although the writer wonders whether the ultimate, even the proximate, effects of science and technology on society are predictable, at the same time he is attracted to the possibility (e.g., the effects of the internal com-

---

[2]Lester B. Lave and Eugene P. Seskin, "Air Pollution and Human Health," *Science,* vol. 169, August 21, 1970, pp. 723-33.

[3]John Walsh, *Science,* vol. 165, September 5, 1969, p. 994.

bustion engine in the year 1910; of inorganic chemicals a decade later; of atomic energy; of jet propulsion; and, now of biochemistry and molecular biology). Such an instrumentality should (1) be equipped to identify new problems of technological significance to society; in time (2) to act on them before they become acute; while (3) receiving continuous feedback from those administering the necessary social constraints about the effects, good and bad; and with the purpose, among other things, of (4) reducing the lead time between the identification of a problem and taking the necessary steps to deal with it.

How many and which of the steps which have been already taken to reform our governmental institutions meet these specifications? On first examination, such agencies as AEC, NASA, even FCC, FPC, and DOT, all of which were taken in response to rapid technological change, fail to meet them. They are agencies to be monitored rather than to perform the monitoring function. Other steps, such as the NSF, help to gain knowledge, hence are designed to perform a different function. Other changes, in the form of additions, have been made in the executive branch, such as the Office of Science and Technology (OST) in the executive office of the president, headed by a director who is also the science advisor to the president, and a science advisory committee (PSAC). A Federal Council on Science and Technology was set up by executive order in 1959. Potentially, this cluster of institutions could develop into one meeting most of the specifications of our model.

A model on a different scale, indeed, on a higher order of magnitude, is Planet Earth conceived as a spaceship. Recognizing that the life support system carried by manned lunar probes is identical in function to Earth's biosphere, penetrating intellects such as the economist Barbara Ward have suggested that Earth be regarded as a spaceship for purposes of making environmental policy.[4] For purposes of supplying operational criteria for environmental management, Lynton Caldwell, a political scientist, adopts

---

[4]Barbara Ward, *Spaceship Earth*, New York: Columbia University Press, 1966.

the spaceship earth concept.[5] Our model, and the recommendations to approximate it, is advanced in the same spirit and embraces the same ethical and management concepts.[6]

In 1970 under authority granted by Congress, President Nixon set up two new agencies, the Environmental Protection Agency (known as EPA), an independent agency, and the National Oceanic and Atmospheric Administration, in the Department of Commerce. The former has responsibility for control of air and water pollution and solid wastes. It includes such units as the Federal Water Quality Administration, transferred from the Department of the Interior. The latter includes the Environmental Science Services Administration (ESSA) and other activities such as the marine fisheries and marine mining programs transferred from Interior and the sea grant program from the National Science Foundation. From HEW's Bureau of Radiological Health and from the Federal Radiation Council EPA takes over the fixing of standards of radioactivity, particularly with the president's approval, determining the amount of radiation to which a person may safely be exposed.

Other structural changes are forecast. In a proposal unprecedented in scope President Nixon asked the 92d Congress to set up four new cabinet departments—natural resources, human resources, economic development, and community development and abolishing seven existing departments by transferring their personnel, programs and authority to the new departments. Congress having failed to act, Mr. Nixon effected as much of the reorganization as possible by executive order.

In addition to these structural changes, existing and prospective, in the federal bureaucracy, Congress has taken steps to modernize itself. As part of the reform undertaken in the Legislative Reorganization Act (1946) both houses provided for new standing legislative committees, a House Committee on Science and

---

[5]Lynton Caldwell, *Environment. A Challenge to Modern Society,* Garden City, New York: Doubleday & Company, Inc., 1971.

[6]See chapter 5, for the recommended national priorities program to solve a very complex socio-economic-political problem.

Astronautics and a corresponding Senate Committee on Science and Aeronautics. Further changes, mostly procedural, but facilitating the taking of record votes by House members, were made in 1970.

DOUBTFUL EFFECTS OF STRUCTURAL CHANGES IN GOVERNMENT

One may well doubt the beneficial effect on the environment of structural changes in government. A variety of factors—bureaucratic inertia, insider perspective, industry resistance, the ambivalence of scientists and technologists, congressional divisions, confusion incident to reorganization—all these, among others such as administrative sluggishness and private foot-dragging, combine to place almost insuperable barriers in the way of carrying structural reorganizations into effect. A number of straws in the wind may be mentioned.

Item. Only after three-quarters of a century has the federal government taken cautious steps to use the Federal Refuse Act of 1899 to control the pouring of industrial refuse into lakes and streams. When an attempt was made, through the Advisory Council on Federal Reports, to supply the Corps of Engineers with the data necessary to enforce this law, the council, organized, financed, and chosen by industry, was able to delay for seven years the preparation and clearance of the necessary questionnaires for an inventory of industrial wastes entering the nation's waterways. The House Committee on Government Operations reported this in late 1970.[7]

Item. The first federal review of the public land laws in nearly a century produces a report singling out use (growing trees for timber) from among other values (wilderness preservation, wildlife, erosion control, mining, conservation, water power, grazing, recreation) as the principle to be followed in future administration of the public lands.[8] Thus, Congress tends to be use-oriented,

---

[7]*The Washington Post,* December 12, 1970.

[8]Report of the Public Land Law Review Commission P.L. 88-606 (1964).

and single-use oriented at that: "In spite of all the fuss about multiple use of land and resources," William E. Towell has observed, "we all tend to be single-use oriented."[9]

Not until the 91st Congress (1969-70) did the legislature exercise its preemptive powers (as against the states) in the matter of authorizing the fixing and enforcement of national air quality standards for ten major polluting substances (1970 Clean Air Act).

By 1970 the federal government and a few states had acknowledged the citizen's environmental rights and had given him standing in court to sue polluters, both private companies and government agencies. But these rights wax and wane according to the prevailing winds of doctrine and propaganda in Congress and in the Courts.

Item. For abating air and water pollution the prevailing forces in Congress succeeded in limiting the federal role to fixing regulatory standards with backup authority and authorizing funds for helping the states to meet these standards while leaving administration to the states. As already noted, forty-one of the states have regulatory boards, either air or water or in combination, on which major pollutant sources—industry, agriculture, county and city governments—are represented. In such a situation, while some form of advisory role for private and public polluters is probably desirable and necessary, conflict of interests is always a possibility. To place their representatives on state regulatory bodies would appear to be a case of "asking the foxes to keep guard over the chickens." When such representation is combined with industrial domination of the advisory council on federal reports (see above), the roadblocks to prompt and effective pollution abatement appear to be numerous and formidable.

Some of these roadblocks had appeared by 1972. States were encountering difficulties in preparing and submitting abatement plans to the EPA administrator by the deadlines set by him.

---

[9]Quoted in *American Forests*, vol. 75, no. 12, December 1969, p. 15. Mr. Towell is executive vice president of the American Forestry Association.

Setting up and staffing a field organization by EPA took longer than expected. Jurisdictional problems between EPA Washington and field officials had appeared. Some state officials were not receiving from Washington the assistance expected, with strained relations the result. In Washington itself reports were prevalent, later denied, that the Office of Budget and Management and the Council on Environmental Quality, both in the Executive Office of the President, differed with EPA over the latter's responsibilities and duties under the Clean Air Act and amendments. No doubt some of these problems cannot be avoided when existing statutory authority of federal departments and agencies is transferred between departments; when governmental reorganization occurs, as in 1970; and when new authority is given departments and agencies, as Congress did in the 1970 Clean Air Act. But when laws as complicated as pollution-abatement laws are adopted over the opposition, or, at best, the reluctant acquiesence, of large parts of public opinion, as the Clean Air Act and amendments were, not all of the roadblocks to efficient administration and enforcement can be attributed to organizational difficulties. Old attitudes toward use of resources; old attitudes toward industry-government relations; old attitudes toward the division of authority between federal and state governments—all these and more yield slowly to the demands of a new environmental ethic.

Administration and enforcement of federal legislation authorizing and funding regulation of the use of insecticides, herbicides, and fungicides illustrate the situation. Only under the major impact of Rachel Carson's *Silent Spring* (1962) and the furor over environmental pollution created by indiscriminate use of chlorinated hydrocarbons as pesticides were moves made to rationalize, in some degree, the federal government's stance toward this source of pollution. Despite a report from the President's Science Advisory Committee (PSAC) requested by President Kennedy and agreeing generally with *Silent Spring's* findings and recommending "orderly reduction in the use of persistent pesticides" with elimination as the goal, only in 1964 was the Federal Committee on Pest Control (FCPC) created. An earlier advisory review board, set up in 1961, was abolished at the same time. It proved ineffective because it reflected the conflicting responsibilities for the control of pesticide use. The FCPC has as its charter, not an executive order, but a kind of treaty signed by the secretaries of the four departments represented on it—Agriculture, Defense, Health, Education and Welfare, and Interior.

Eight years later advocates of a complete ban on DDT had still not reached their goal. There are many reasons. For one, the secretary of agriculture opposed and resisted the petitions and court orders obtained by a group of environmentalists on the grounds that substitutes might be even more dangerous and that scientific evidence does not show DDT to be an imminent hazard to human health. Second, the pesticide industry has effectively blocked, by resort to the courts, the cancellation proceedings of the registration of DDT for use on many kinds of vegetables and fruits, forest trees, lumber, livestock, and in buildings. Third, speedy action on the phaseout of DDT for all but essential purposes (cotton growing accounts for more than three-quarters of the DDT used in the U.S.) announced by the administration as early as November 20, 1969, was hindered by governmental reorganization and by the transfer in 1970 of registration and cancellation responsibilities from the secretary of agriculture to the administrator of the Environmental Protection Agency. Several personnel changes in EPA may help explain the delay, if not the reluctance, of EPA to carry out the federal Court of Appeals order to ban DDT. It was announced on May 10, 1971, and later confirmed, that "EPA's insecticides chief, Dr. Raymond E. Johnson, an impeccable bureaucrat with deep knowledge of the effect of insecticides on wildlife, is to be replaced by William M. Upholt, secretary of the do-nothing Federal Committee on Pest Control and a critic of recourse to the courts in settling scientific controversies."[10] Finally, exploitation by DDT's largest manufacturer of the views of Dr. Norman E. Borlaug, a plant geneticist and Nobel Prize winner, on DDT as an insecticide, among other things, has not helped the EPA in its efforts to phase out DDT for all but essential purposes nor environmentalists in their efforts to ban it.[11]

---

[10]*Audubon,* vol. 73, no. 4, July 1971, p. 93.

[11]See chapter 2 for more information on this incident.

A key question in our inquiry into the ways in, by, and through which technology is related to environmental policy is, what pattern, procedure, or device should be employed to deal with scientific controversies?

PROPOSALS FOR IMPROVING
TECHNOLOGICAL ASSESSMENT

In addition to changes already made by the Congress and the president in the government's structure and functions numerous proposals have come from individuals and groups and are still to be acted upon.

1. Proposals for technological assessment boards have been made by a National Academy of Science (NAS) panel at the request of the House Science and Astronautics Committee;[12] by former Representative Emilio Q. Daddario, chairman of the subcommittee on science, research, and development;[13] and from Leo Marx, a professor of English and American Studies at Amherst College.[14] Both the legislative and executive branches would be supplied with new boards by the NAS proposal. The Daddario proposal would be limited to Congress, supplying it with a new staff arm modelled after the General Accounting Office. The Marx proposal, on the other hand, is for a panel established by the American Association for the Advancement of Science (AAAS), independent of the government, to take advantage of the apparent convergence of our scientific and literary views and our expansionary life-style.

In the herbicide 2,4,5-T decision (1971), (discussed later), further evidence is available to illustrate the many contexts characteristic of present-day attempts by government to protect the public health from further deterioration of the environment.

Before proceeding to look at proposals for improving techno-

---

[12]*Technology: Processes of Assessment and Choice*. Report of the National Academy of Sciences. Committee on Science and Astronautics, U.S. House of Representatives, July, 1969.

[13]See report in *The New York Times*, March 25, 1970, p. 23.

[14]Leo Marx, "American Institutions and Ecological Ideals, *Science,* vol. 170, November 27, 1970, pp. 945-51.

logical assessment, let us refer again to one structural change or organizational device, the wartime OSRD, which may repay examining in more detail. As already noted, OSRD's wartime success in research and development in meeting specific needs may well bear close scrutiny for meeting the present environmental crisis. This is done in chapter 5.

2. Quite different are two suggestions for dealing with the so-called information explosion: one, improving the accessibilty of existing and new information to scientists through coordination of individual information systems into an effective information network;[15] and two, enabling the individual citizen, as he goes to the polls, to obtain ecological wisdom from a data bank, established, kept continually up to date, and serviced by government.[16]

3. Far-reaching in its implications is the third type of proposal, the policing of research in biology, involving the creation of national and international machinery for this purpose.[17] The researcher making the proposal, Dr. James F. Danielli, is quoted as saying that "within a century we shall probably be able to synthesize artificially any biological system or entity"; that new types of organisms might produce harmful strains that would plague the world if they escaped; and that researchers in this field cannot be expected to police themselves. In the United States the National Academy of Science or the National Science Foundation might be asked to assume the job; outside the U.S. some unspecified inter-international agency might be designated. A "watchdog over science" has also been proposed by a group of concerned scientists meeting in New York.[18] And the future envisaged by Danielli seems to be approaching rapidly, if it has not already arrived.[19]

---

[15]Dale B. Baker, "Communication or Chaos?" *Science,* vol. 169, August 21, 1970, pp. 739-42.

[16]Harvey Wheeler, "The Politics of Ecology," *Saturday Review,* March 7, 1970, pp. 51-52, 62-64.

[17]"Life-Synthesis Pioneer Urges Policing of Research," *The New York Times,* December 8, 1970, p. 32.

[18]*The New York Times,* December 19, 1971.

[19] Willard Gaylin, "We Have the Awful Knowledge to Make Exact Copies of Human Beings," *The New York Times Magazine,* March 5, 1972.

Organized research in molecular biology already confronts the country with some new questions about basic research and the private enterprise system.[20]

4. Quasi-official agencies are studying technological assessment for official agencies. The Office of Science and Technology is reported as having contracted with MITRE Corporation, a "think tank" located at McLean, Virginia, to prepare a study of technological assessment for it.[21]

TECHNOLOGICAL ASSESSMENT: VALUES AND DEFECTS

Basically, technological assessment involves choices, choices which must be made from among competing alternatives, no one of which is clearly right; under the compulsion of an advocate system favoring producers; and in the absence of any guarantee that the long-range public good will be served. How it works is well illustrated by the long-drawn-out decision on the herbicide, 2,4,5-T.

A powerful weed-killer, 2,4,5-T has been used in large quantities in Vietnam by the military, as well as by commercial farmers, gardeners, and others in the United States. It was first suspected of being teratogenic (tending to cause fetal deformity) in 1966. By then its use had become widespread on food crops, around homes, and near water. Upon first being marketed it had not been proved hazardous to public health, nor had it been proved not to be hazardous. Under federal law no such proof was required. Since 1966 there has been a growing consensus among scientists that 2,4,5-T and its dioxin contaminant are both toxic and teratogenic at some dose level. On April 15, 1970, its use on crops, near water and around the home was cancelled, an action challenged

---

[20]"Molecular Biology: Corporate Citizenship and Potential Profit," *Science,* vol. 174, October 15, 1971, pp. 275-76.

[21]NASA might well be considered as a model for dealing with environmental degradation on a crash basis. See Foy D. Kohler and Dodd L. Harvey, "Administering and Managing the U.S. and Soviet Space Programs," *Science,* vol. 169, September 11, 1970, pp. 1049-56.

by Dow Chemical Company and Hercules Incorporated, who asked, as was their right under the law, for the setting up of a scientific advisory committee to review the directive as it applied to crops. Unlike injunction proceedings in equity, in which a court can issue a directive to prevent harm from being done, pending a hearing, while the appeals process under the herbicide law is going on the use of 2,4,5-T on crops continues. Set up by the Department of Agriculture, which then administered the law, on nomination of the National Academy of Science-National Research Council, the advisory committee in May 1971 recommended to the EPA administrator, to whom enforcement of the law had been transferred in November 1970, that the restrictions on all uses of 2,4,5-T be lifted. Three months later EPA Administrator William D. Ruckelshaus rejected the recommendation when he announced that the cancellation order as it pertained to the use of 2,4,5-T on crops would be maintained, at least until a public hearing, the final stage of the appeals process, was held.

However, the EPA administrator's decision was reached only after months of delay during which numerous questionable features of the overall process were revealed. The Scientific Advisory Committee's report was leaked to the public, deliberately, it seems, in order to publicize the division within the Committee. In addition, the dissenting member's report was likewise made public. Two environmentally-oriented groups, the Committee for Environmental Information and the Center for the Study of Responsive Law, were instrumental in breaking the code of secrecy surrounding such committee reports, a practice long sanctioned while the Department of Agriculture was the enforcing agency. Reputable scientists both in and out of government backed the dissenting member's criticism of the committee report. All this took place during the administrative birth pains of the EPA, an independent agency with 8,500 employees and an operational budget for fiscal year 1972 of 2.4 billion dollars. And the process, the bureaucratic red tape, is not yet unwound. After the public hearing, referred to by Administrator Ruckelshaus, comes the administrator's decision, which is itself subject to review by the courts.

Meanwhile, some major questions arising out of the 2,4,5-T

decision may be recapitulated: How should scientific advisory committees be constituted? What criteria should be used in the selection of members? Should nominees, otherwise qualified, be barred by financial interests in the producer companies concerned? Who should nominate committee members? The NAS-NRC has asked to be relieved of this responsibility, now placed on it by law. If not the NAS-NRC then who? Should committee membership be balanced between those members environmentally-oriented and those not? Shouldn't all public business be conducted in public meetings? Still other questions are in need of answer. Should poisonous chemicals such as 2,4,5-T be marketed before their safety is determined? By whom and how is safety to be determined? Is it in the public interest to interrupt the scientific decision-making process by transferring responsibility from an experienced department or agency to a new, hence inexperienced, agency in the interests of rational government organization? The basic question is, perhaps, should the public be exposed to 2,4,5-T or any such product before or after the toxic and teratogenic dosages have been determined?[22]

Probably the single most serious handicap of all proposals for technological assessment is that it is almost entirely alien to the American ethos. The philosophy and methods of technological assessment are grounded in the tacit assumption that persons, both legal and natural, are willing to face the consequences of their actions. Historically, this is not in accord with American beliefs and values. Historically, risk-taking rather than consequence-facing, epitomizes the American outlook. Nor is technological assessment much closer now, although the reemergence of ecology (America has been warned of the environmental consequences of uncontrolled risk-taking for at least a century), and the heightened consciousness of youth and of many adults to the unwholesomeness of our surroundings, may hold some promise for a change. Also, what may be called the technological mystique, that is, the

---

[22]Arthur H. Westing, "Ecocide in Indochina," *Natural History,* vol. 80, no. 3, March 1971, pp. 56-61; *Science,* vol. 173, July 23 and August 13, 1971, pp. 610-15.

obsession that all our ills can be solved by applied science, is so prevalent that some means must be found for harnessing the dynamism of professionals and the faith of laymen in the potential of research and development to improving the environment. Moreover, so ingrained in the economy and the government is the belief in and practice of government support for research and development that the strength of this belief and the extensive practice based upon it cannot be dampened, let alone choked off, even if it were desirable to do so. This suggests the value of examining the World War II Office of Scientific Research and Development as a prototype for identifying and solving the currently intractable problems traceable to a galloping technology.

BEYOND TECHNOLOGICAL ASSESSMENT

Granted the urgent need for improved means of technological assessment improved means by themselves will not do the job. In the absence of a reversal of outlook from the present to the future all these reforms will produce little difference in our ability to master the detrimental effects of technology upon society at large.

What is needed more than technological assessment, in the sense of better governmental institutions and improved procedures, is a resetting of our sights for the America of the future. What form this should take should be put at the top of our national agenda. Some envisage it as a statement of national goals. One such statement was prepared, apparently at President Nixon's request, released in midsummer 1970, and was greeted with almost complete silence.[23] Others, more practical, would reset our sights by revising policies toward the public lands. Insofar as it deals with natural resources on public lands, the Public Land Law Review Commission's (Aspinall) Report leaves much to be desired, although it could conceivably, in the course of its passage through Congress, be reworked into the kind of policy declaration required. Still to

---

[23]"Toward Balanced Growth: Quantity with Quality," report of the National Goals Research Staff (U.S. Government Printing Office, Washington, D.C., July 4, 1970).

be acted upon by Congress is President Nixon's proposal for federal assistance to the states in drawing up new land-use laws.

Others envision a pioneering bill of environmental rights, building on the United Nations' Declaration of Human Rights and, more particularly, the U.N.'s Covenants of Political and of Social and Economic Rights. The 1972 Stockholm Conference on the Human Environment adopted a Declaration on the Human Environment, among other things, for presentation to the UN General Assembly and its members for implementation. Still another group sees antiquated congressional procedures as the roadblock. Changes made by the 91st Congress are good but do not go far enough. Still others wonder whether we might not resuscitate old-fashioned faith in observing the past as a guide to the future, relying less on computer printouts and more on records already available. For example, Louis H. Bean punctures the popular fallacy that weather records, as a basis of predicting crops and floods, are random in nature, hence of little value, when in fact they show definite patterns over the past fifty to a hundred years and can be used to predict crop yields and floods both here and abroad with remarkable accuracy.[24] The moral is not that we should use modern technology less in predicting the effects of invention and discovery on humans and their environment but that, in our haste to apply scientific innovations, we not be led to abandon time-tested qualities of foresight, prudence, and adherence to humane standards.

But a return to old standards is not enough. There is great need for a willingness to reexamine some of our most cherished myths. Two major ones may be mentioned here. One is the feasibility of unlimited national growth, the mystique that finite resources on a finite spaceship, Earth, will suffice for fueling a constantly expanding gross national product.[25] Nationally, quantity has been accepted as the highest value, thus rejecting quality.

---

[24]Louis H. Bean, *The Art of Forecasting,* New York: Random House, 1969.

[25] J. Alam Wager, "Growth versus the Quality of Life," *Science,* vol. 168, June 5, 1970, pp. 1179-84 at 1179.

Today those questioning the wisdom of continued seeking of a constantly-rising GNP are voices mostly unheard and certainly unheeded. Despite the statistics showing 1970 to be the first in many years without an increase in Gross National Product, there are as yet few who view the financial burdens of the cities in terms of a better distribution of a stable national product rather than in a bad distribution of a rising national product. Use of the federal taxing power to reduce the population of the cities, as one means of stopping rural-urban migration, and, consequently, of easing the welfare load, might conceivably be a factor in flattening out the GNP curve. If the *National Observer* had its way, lessening the welfare component of the cities' financial burden in this way would appear to be worth the cost. Shelve revenue-sharing, the *Observer* proposed, institute standardized welfare regulations throughout the nation and simultaneously reduce the population of the cities as a matter of federal policy by offering tax incentives to businesses that set up facilities away from the city centers.[26] A smaller welfare load on the cities and states would not be the only result. Numerous environmental problems would yield more readily to solution if the population of the cities were to be reduced, thus signifying the breaking away of our country from some outmoded economic thinking.

The other myth may be disposed of quickly, the myth that the federal budget is an accurate statement of our national priorities. Two points are relevant: one, whether consciously or not, the assumptions made by the president and his advisers in drawing up the federal budget are of a partisan political character, not very well veiled;[27] and two, despite the statutory authority of the chief executive to use the budget as a statement of national priorities, the rivalry between the Congress and the chief executive built into our constitutional system guarantees congressional modification, if not alterations, of these priorities. Only in the grossest sense may the budget be regarded as reflecting our national values.

---

[26]*The National Observer*, February 1, 1971, p. 10.

[27]See, for example, "Nixon's Budget Mystery" by Leonard S. Silk, *The New York Times*, February 3, 1971, pp. 1, 49.

In our struggle to reset our sights more on quality and less on quantity we may be aided by the slowing-down of the revolution caused by science-based technology. Although what Boulding calls "galloping science" would appear to show few signs of slowing down, he among others believes this to be the case. In an article in the *Bulletin of the Atomic Scientists* he argues that any natural phenomenon, such as the scientific subculture, which develops exponentially cannot continue to do so indefinitely, and that it does not do so very long. For nearly three hundred years, he argues, science has been developing at this phenomenal rate. It cannot go on much longer although in certain fields, such as molecular biology, the portents are disquieting. In any event, as with population growth in the United States, so with scientific growth: the built-in kinetic factors in science guarantee for the immediate future continued technological growth, although not at the same breakneck pace.[28]

Since policy-making in the United States is divided between the Congress and the chief executive, with the president as chief proposer and Congress the chief disposer of proposals, both branches should have attached to them, in a staff capacity, a technological assessment board or its equivalent. As regards Congress this might well be the present General Accounting Office, (GAO), with its authority properly modified to include forecasting of future effects of technological change, and tied to the Library of Congress Legislative Reference Service and staffs of the standing congressional committees. Because of its accountability directly to the Congress; of the long-term (fifteen-year) tenure of its head, the Comptroller General; of the high standing it holds with the Congress due to its expertise in determining how well the executive agencies and departments have carried out the intent of Congress—the GAO might well be asked to add the function of technological assessment to its present functions.

As regards the chief executive and the executive branch it is difficult to imagine how any agency not in the executive office of

---

[28]Kenneth W. Boulding, "The End Is in Sight for Galloping Science," excerpted in *The Washington Post*, September 6, 1970.

the president could acquire the prestige necessary to discharge the multi-functional assignments which would be incumbent on a technological assessment board. It is likewise difficult to envisage an agency within the executive office doing what would be required of a technological assessment board and retaining the confidence of the chief executive. As already indicated, such a board would need to know not only what is going on in scientists' heads within the government but also in industrial laboratories in the country at large. As a guideline the "need to know" criterion should override every other consideration if the board is to be able to carry out its function of general staff in the fight against environmental pollution. Such a fanciful notion is never likely to be realized in practice. It makes abundantly clear, however, that the head of the technological assessment board, like an army chief of staff, can be denied no information necessary to the discharge of his duties. And, like the comptroller general, he should hold his appointment for a long term of years, so that, like him, he would be insulated from the political winds of the moment.

To illustrate, heightened sensitivity among government officials would, using the analogy of infrared sensors, trigger a mechanism in the federal bureaucracy warning that some stage in technological assessment is being bypassed, overlooked, ignored, or underassessed. In a bureaucracy of the size and complexity of the federal government's executive branch circulation of new information among all agencies "needing and/or should know" is not automatic; it has to be provided for and then monitored. Even such an important need as a central point for gathering and publishing legally binding regulations issued under rule-making authority delegated by Congress was met only in the 1930s, with the beginning of the Federal Register.[29] Something more than a require-

---

[29]In the days of the National Industrial Recovery Act (NIRA, 1934-36), regulations covering wages, hours, and working conditions and having the force of law were being issued by Code Authorities but the body of these regulations was nowhere being collated. With the establishment of the Federal Register in 1936 all such regulations and all other regulations issued by rule-making authorities had to be published therein. The Federal Register has outlived NIRA, which expired in 1936, when the Supreme Court in the Schechter case struck down the code-making authority.

ment to publish is needed if technological assessment is to have significant meaning. If a simple requirement to publish was necessary in the simpler days before World War II, how much more is a multidimensional assessment mechanism needed today.

In the Environmental Policy Act Congress enacted a requirement that any agency proposing legislation or action significantly affecting the evironment prepare an environmental impact statement and circulate it to the EQC, other concerned federal agencies, and the public. So far, however, no clear understanding as to how this requirement should be implemented has been reached. On this crucial point the law is vague. Demand for clarification is already being made and it may well be that Congress will amend it. Within the administration, however, Secretary of the Interior Morton is reported in 1972 to have opposed any change in the law.[30]

Meanwhile, the chairman of the Council on Environmental Quality interprets the law as requiring only the final, circulated and commented-on statement be made public. On the other hand is the opinion of the president of the Friends of the Earth, who claims that such an application of the law "threatens to smother in its cradle" the National Environmental Policy Act itself.[31]

Similar if not greater potential danger to the public occurs when no agency has clear-cut authority to test a new product, such as enzyme detergents, before it is placed on the market, and when industry and extra-industry and government scientists clash over whether the product, when used on a mass basis over time, will be damaging to the public health.

Drawing up a trial balance would appear to show that, although at this time the results are by no means in from the steps which have been taken to meet the environmental crisis, still it would seem that none of the steps taken, either singly or together, will do the job. Structural changes, procedural reforms, even formal moves to recognize the need for the new function of technological assessment, and its improvement—none seem adequate. An ex-

---

[30]*The New York Times,* March 10, 1972.

[31]David Brower, "Who Needs the Alaska Pipeline?" *The New York Times,* February 5, 1971.

pert, a political scientist, say, one who at the same time is a generalist versed in environmental studies, might suggest reorganizing our governmental structure along river-basin lines, updating John Wesley Powell's General Plan of the 1880s through constitutional reform to make the Constitution over in a modern image. In the light of even more drastic proposals for redistributing governmental power, such an idea as a combination river basin-regional principle for governmental reorganization does not appear fanciful. Still, advocates of such a scheme could probably not, at this time, generate much political support. Hence while envisaging it as the ultimate solution to the environmental problem, a proximate solution based on a priority program somewhat less visionary. in content is proposed (chapter 5).

4.
Energy Needs, Technology, and Environmental Policy

The problems of technological assessment as they relate to environmental policy are rolled together and are inextricably intertwined in the current search for a national energy policy.

To meet projected energy demands into the 1980s, policy is being shaped in a process which is largely in the hands of private interests. It favors atomic energy, oil and gas with greater dependence on overseas sources, deemphasizes coal, and thus far relies little on conservation and on research and development of new sources. While a partner in the process, government is not doing as much as it could. For nearly half a century tax policy has favored producers of nonrenewable resources, mostly oil, through depletion allowances. In the past few years and growing out of World War II development of atomic energy for military purposes, government has been promoting atomic energy for peaceful purposes, especially research on some of the earlier types of atomic reactors for electric power, as well as joint research and development with industry on advanced reactor types. Beyond these activities government has not done much. It is still in the "tooling-up" stage as regards abatement of air and water pollution. As for conservation of scarce supplies of energy, government is still in an even earlier report and exhortation stage.

Demands for electrical energy are doubling every ten years.[1]

1"Electric Power: An Environmental Dilemma," *Sierra Club Bulletin,* February 1972, p. 16.

(Note this is demands, not needs). By 1990 natural gas needs, projected from 1966, are expected to nearly triple. Between 1970 and 1972 increased domestic demand for crude oil and unfinished petroleum accelerated so as to require raising the oil import quota four times. In the domestic gas industry government policy makes the importing of natural gas in liquid form from overseas more profitable than increasing production from domestic sources. Federal government support of research and development of alternative sources of power—fuel cells (except for space probes), solar energy, geothermal energy, atomic fusion, magenetohydrodynamics —lags. Most federal R&D funds on power sources other than fossil fuels go to support industry-government construction of new atomic energy plants—the liquid metal fast breeder reactor (LMFBR). The funds for developing methods of converting coal into a clean gaseous fuel—coal gasification—are small in comparison.

As for conservation—how the country as a whole could cooperate in reducing energy demand—elaborate studies have been made by the federal government's Office of Emergency Preparedness.[2] But knowledge of the potential savings estimated in these studies is not being promoted. This is not part of present energy policy.

Simultaneously members of Congress and some federal officials testifying before congressional committees spurn suggestions of other witnesses, for example, Admiral Hyman G. Rickover, that growth rates of energy demands be slowed, that this be accomplished by taxation, that wasteful habits be reformed, and that conservation measures be instituted on a nationwide scale.[3]

So far as meeting existing and projected energy needs is con-

---

[2]Executive Office of the President, Office of Emergency Preparedness, *The Potential for Energy Conservation,* a staff study, Washington, D.C.: U.S. Government Printing Office, October 1972.

[3]Hearings before the House Committee on Interior and Insular Affairs, testimony of Admiral Hyman G. Rickover; (then) Secretary of the Treasury John B. Connally; Secretary of the Interior Rogers C. B. Morton; Federal Power Commission Chairman John W. Nassikas, reported in *The New York Times,* April 23, 1972.

cerned, both private and governmental efforts to shape policy run into environmental problems because our experience with technological assessment in this context is limited, contradictory, and complicated. It is limited because the statute under which this experience is being gained was enacted only in 1969. It is contradictory because there is little consensus among the experts assessing the effects of technology on the enviroment. It is complicated because administrative officials disagree on how the National Environmental Policy Act (NEPA) of 1969 should be administered; because the bureaucracy, the courts, and the public have different ideas about what the Act calls for; and because while the goal of a better environment is agreed upon, the means of reaching this goal—the policies, guidelines, and standards required to reach it—provoke wide disagreement.

The National Environmental Policy Act came into effect January 1, 1970. It has been called "one of the most revolutionary bills in the history of this country."[4] In the nature of the situation the oil and gas and electric-generating and manufacturing industries with the tacit support, if not encouragement, of federal officials, were eager not to allow NEPA to stand in the way of expanding supplies and facilities to meet increased demand. And neither industry nor government rushed to acquire extensive experience in bringing energy policy into line with so revolutionary a change in the nation's stated course.

Nevertheless, despite industry's attitude, an extensive body of experience was forced upon both industry and government by outside intervention. Through numerous agencies the public intervened, mostly by using the courts to spell out how the requirements of NEPA applied to the oil and gas and electric power service industries. "In the first 26 months following enactment of the National Environmental Policy Act there were more than 75 judicial decisions construing NEPA," Senator Henry Jackson is

---

[4]John B. Calhoun of the Laboratory of Psychology of the National Institute of Mental Health, quoted in Lynton Keith Caldwell, *Environment: A Challenge to Modern Society,* Garden City, New York: Doubleday & Co., Inc., p. 227.

quoted as saying in a foreword to a Library of Congress study.[5] In other words, along with other industries whose activities were actually or potentially affecting the environment, the oil and gas and electric power service industries with their advocates and allies in the government had to explain in court the extent to which their programs of expansion conformed or failed to conform to NEPA. The large number of cases in such a short space of time is a measure of the disagreement among experts and laymen on the effects of technology, existing and prospective, on the environment.

Another aspect of the same situation is the complicated character of the process of trying to meet increasing energy demands while complying with the requirements of NEPA. As readers are already aware, even before NEPA the oil and gas and electric-generating companies had to comply with various regulations, both federal and state, as conditions of construction and operation. To build and operate atomic reactors investor-owned utilities must receive licenses from the Atomic Energy Commission. The building of gas and oil transmission lines comes under the jurisdiction of the Interior Department, likewise the drilling and operation of offshore oil wells by the oil industry also licensing of hydroelectric plants and fixing the price of natural gas. Import of oil from foreign sources is governed by quotas fixed by the federal government. While there are exceptions (Tennessee Valley Authority, Bureau of Reclamation projects such as Bonneville and Hoover Dam, and the consumer-owned Rural Electrification Administration), by and large the electric-generating industry and the oil and gas industry operated under a body of federal law enacted by Congress and administered by the federal bureaucracy, so far as interstate and foreign operations are concerned, and under state legislation and administration for intrastate operations.

Now, with the enactment of the National Environmental Policy Act, far-reaching requirements were added to those already embodied in public utility and gas and oil regulatory legislation. In

---

[5]Warren Donnelly, *Effect of Calvert Cliffs and Other Court Decisions Upon Nuclear Power in the United States,* Washington, D. C.: U.S. Government Printing Office, 1972.

the light of the experience already acquired possibly the most important was that calling for environmental impact statements on "proposals for legislation and other major Federal action significantly affecting the quality of the human environment. . ." (Public Law 91-190 91st Congress Sec. 102 [C] [i]). For the electric power service industry probably the most sweeping decision was in a suit brought by three environmental groups against the AEC for failure to apply NEPA provisions in its consideration of the licensing of the Calvert Cliffs nuclear generating plant under construction on Chesapeake Bay. In 1971 the U.S. Court of Appeals for the District of Columbia found for the plaintiffs in a decision which criticized the AEC for its "crabbed interpretation" of NEPA which "makes a mockery of the Act." Although the decision was applicable to eighty-eight generating units then under construction or in operation, the AEC decided not to appeal the decision. Instead it heeded the Court's admonitions; drastically revised its rules for licensing power reactors, fuel processing facilities, and uranium mills; and, under a new commission chairman, James R. Schlesinger, adopted a new policy of refereeing the nuclear power industry instead of promoting it. Seventeen months later, in December 1972, Mr. Schlesinger was nominated by President Nixon to be director of the Central Intelligence Agency.

NEPA not only requires environmental impact statements but also detailed statements of "alternatives to the proposed action." Noting, in the Calvert Cliffs case, that NEPA told federal agencies that environmental protection deserved an equal footing with the promotion and regulation of industry, the Court criticized AEC for, in effect, ignoring this requirement of the Act. As a matter of policy Congress intends AEC and other federal agencies to consider alternatives on a case-by-case basis. But the question might well be raised why new industries could not be built as alternatives or as supplements to existing industries in meeting the energy crisis.

A part of the nation's energy needs for domestic heating and lighting might well be met by a new fuel cell industry.[6] New methods of coal gasification, if widely adopted, would use coal,

---

[6]*The Washington Post,* November 12, 1972.

the country's most plentiful fuel, thus releasing us from undue dependence on foreign sources and with far less pollution to the environment than from existing methods.[7]

As part of the National Science Foundation's Research Applied to National Needs Program, a report of a conference on geothermal energy prepared by former Interior Secretary Walter J. Hickel concludes that a ten-year research and development program on geothermal energy could be producing 132,000 megawatts of electricity by 1985, amounting to about 38 percent of present United States capacity. The estimated cost would be about $685 million, about three-fifths of the estimated annual cost of needed R&D in the electric-generating industry.[8]

A Washington journalist offers an interpretation of the recently discovered electricity crisis and offers a permanent solution. As the country's biggest industry supplying a vital service under monopoly conditions, the electric power industry, he says, can manipulate public opinion and legislators at will. To counteract the industry's brinkmanship he proposes a national power grid, interconnecting the nation's generating facilities so as to be able to meet any power supply emergency anywhere anytime, thus breaking the grip of local autonomous generating units on the country's one essential commodity.[9]

If economic factors (high comparative costs) make oil extraction from tar sands and oil shales unprofitable, is this not a compelling reason for expanding federal funds for research and development of new methods? It would seem so, particularly since most of the deposits are on land already owned by the United States.

A comprehensive report on electric power and the environment and, incidentally on energy needs, technology, and environmental policy appeared in mid-1970, when the Energy Policy Staff of the

---

[7]"Ready to Turn Coal into Gas," *Business Week,* no. 2225, April 22, 1972, pp. 44 ff.; "A 'New' Fuel," *The New York Times,* February 27, 1972.

[8]*The New York Times,* December 11, 1972.

[9]Robert Sherrill, "Power Play," *Playboy,* vol. 18, no. 5 (May 1971), pp. 113-14, 224-29.

Office of Science and Technology, at that time headed by S. David Freeman, responded to a request of Dr. Lee A. Dubridge, then OST director, for a "program for resolving the environmental problems that have emerged in the siting of steam electric power plants and extra high voltage transmission lines.[10] The report proposes a four-part program:

1. long-range (ten years in advance of construction) planning of utility expansions on a regional basis

2. participation in such planning by the environmental protection agencies with notice to the public at least five years in advance of construction

3. preconstruction review and approval of all new large power facilities by a public agency at the state or regional level or by the federal government if the states fail to act

4. an expanded program of research and development aimed at better pollution controls, underground high-voltage power lines, improved generation techniques, and advanced siting approaches so as to minimize the environmental problems inherent in existing technology[11]

Proposed legislation to implement the proposals was forwarded with the report.

In its recommendations the report adheres closely to the terms of reference imposed by the OST director. Before arriving at its conclusions, however, the Energy Policy Staff necessarily paid some attention to other parts of the larger matter of energy needs, technology, and environmental policy. While the staff dealt thoroughly with the immediate problems of siting power plants with

---

[10]*Electric Power and the Environment.* A Report Sponsored by The Energy Policy Staff, Office of Science and Technology. In cooperation with Atomic Energy Commission, Department of Health, Education and Welfare, Department of the Interior, Federal Power Commission, Rural Electrification Administration, Tennessee Valley Authority, and the Council on Environmental Quality. Washington, D.C.: U.S. Government Printing Office, August 1970.

[11]In his 1971 special message to Congress on the environment President Nixon included a power plant siting bill. Congress has not yet acted on this proposal.

minimum damage to the environment, at the same time it did not overlook considerations affecting the growth of electricity consumption. In addition its chapter on research and development (VI) throws some light (as of 1970) on the staff's sense of priority research and development pertinent to power plant siting. Roughly a year later the place of these R&D projects in the larger picture under consideration here came to light in President Nixon's $3 billion ten-year package of new fuel policies and research for clean energy. While the president offered something to the environmentalists, and to the oil and gas (fossil fuel) industry, $2 billion of the $3 billion which was requested from Congress would go to the nuclear power industry in the form of research and development of a liquid metal fast breeder reactor by 1980.

In expounding the place of electric power in the different but not necessarily larger picture of energy needs, technology, and the environment, we need to keep reminding ourselves of two things. First, the electric power industry is the largest in the country; and, second, in the shaping of energy policy a large part of the decision process is in the hands of private interests. So far as the electric power industry is concerned, although the country's largest, it is atomized. It is composed of 3,500 individual entities loosely gathered into four institutional groups: the manufacturing group, and the three generating and distributing groups—investor owned, consumer-owned, and the public power association. The other private interests in the energy field are, of course, the oil and gas companies. The oil and gas industry is nowhere near as atomized as the electric power industry. Neither is it a trust or monopoly, as it once was two generations ago. Economists have a word for it—oligopoly, an industry in which two or three or a small number of the units comprising the industry exercise a predominant influence if not control. For this and other reasons the oil and gas industry occupies a predominant position in the energy decision process. Not only is this position so among the private interests in whose hands so much of the energy decision-making process lies. It is also true of the interests whose decisions are so important in the making of *public* policy. In a sense, with about 75 percent of the country's energy coming from oil and gas, a little less than 50 percent from oil and a little over 25 percent from gas, the country literally pays tribute to the predominant position of the oil and gas industry in the making of national energy policy. The country is constantly being reminded of this in the industry's current propaganda pitch: "A country which runs on oil can't afford to run short."

In such a situation it is not hard to understand why new sources of energy, no matter how promising or feasible technologically or less polluting than oil, receive such low standing in the current scale of R&D priorities. In this connection a table, "Priority Research and Development Pertinent to Power Plant Siting," appearing on pages 44 and 45 of the Energy Policy Staff's report, is revealing. For fiscal year 1970 no federal funds were estimated as available for combined power cycles, for fuel cells, or for solar energy and other undeveloped energy sources. Coal gasification is not mentioned. Nor were nonfederal (private) funds expected to be available for these new sources. Some $300,000 federal and $200,000 nonfederal funds were expected for magnetothermodynamics. On the other hand, $30 million federal and $1 million nonfederal funds were expected to support R&D on the controlled thermonuclear reactor (atomic fusion) and $100 million federal and $25 million nonfederal funds for the liquid metal cooled fast breeder reactor.

In President Nixon's 1971 proposals R&D on new sources of energy were likewise low in standing on his list of priorities. However, coal gasification was upgraded, with $20 million federal and $10 million of industry funds budgeted. Other items of promise to the oil and gas industry were an increase in R&D funds for sulfur oxide control technology from $11 million for FY 1971 to $26 million for FY 1972, up from $1.1 million for FY 1970; quickened expansion of offshore oil leasing; and consideration of speeding up the leasing of federal oil shale lands. Open to exploitation by the oil and gas as well as to other sources of capital is the prospect of leasing of more than a million acres of federal lands underlaid by geothermal steam.

But, as already noted, the biggest federal plum in R&D funds in the president's energy plan is the $2 billion commitment asked from Congress for research and development of a single demonstration liquid metal cooled fast breeder nuclear reactor.

The mix of component sub-policies, public and private, in existing energy policy reflects clearly the absence of any underlying values and the yielding to expediency when decision-makers have been confronted by new and un-looked-for situations. A more charitable interpretation is that of Joseph C. Swidler, chairman of the New York State Public Service Commission: "This country had no need for an energy policy ten years ago, or at least most people did not recognize the need for one until late in the last decade."[12] This is another way of saying that so far as energy is concerned we Americans have been making policy by default.

If policy-making in the past has been policy-making by default, neither the present situation nor that of the immediate future holds great promise of improvement. One appraisal among many available is chosen for presentation. In 1972 the staff of the New York State Public Service Commission, a body of experts painfully aware of the accelerating demands for energy, spoke to a much wider constituency than the Commission, when it warned against undue reliance on atomic energy to meet these demands. After stating that "nuclear energy is not the answer to all our problems" and criticizing the large volume of funds, public and private, committed to a demonstration project—the liquid metal fast breeder reactor—one, moreover, incapable of helping greatly in meeting the energy needs of the seventies and with still unsolved technical problems, the staff report endorses fuel cells and coal gasification as technological developments worthy of far more research and development money than is presently being devoted to them. Given the new sulfur standards adopted by Boston, New York City, and New Jersey in the past year, the report warns that coal as an energy source is effectively eliminated from the market place, unless it can be converted to a clean

---

[12]Before the Energy Seminar Program of Oak Ridge National Laboratories and Oak Ridge Associated Universities, April 6, 1972.

burning gaseous fuel by some process such as coal gasification.[13]

In clarifying the uncoordinated forces so characteristic of present-day energy policy-making the Office of Emergency Preparedness (OEP) staff study, "The Potential for Energy Conservation" is helpful. One measure of the magnitude of the potential has been pointed out by OEP's director, George A. Lincoln. In speaking to the (Oil) Producers' Council Colonel Lincoln (Ret.) referred to the wide range of short-, medium-, and long-term energy conservation measures which, given sufficient motivation, could be adopted in the four major energy consumption areas—transportation, residential and commercial, industry, and utilities. Given this motivation, Colonel Lincoln added, a reduction in energy demand by 1980 could be brought about equal to 7.3 million barrels of oil per day, or about two-thirds of the projected oil imports for 1980. "This would be as significant a contribution to balancing the energy equation as adding the production of three or more Prudhoe Bays of our northern Alaskan oil to the supply side . . ." If taken seriously conservation measures could mean a lot in balancing the supply-demand equation.

Why isn't energy policy better coordinated, more consistent internally? The answer lies in the way energy policy is made. The Office of Emergency Preparedness and Colonel Lincoln, its director, can only point out what the country as a whole can do to conserve energy. He can only exhort oil and gas producers and distributors and jobbers and technologists to get behind the conservation measure proposed. While Colonel Lincoln sits on the energy subcommittee of the President's Domestic Council and testifies for OEP before congressional committees, his office can only watch over and coordinate the oil import program. And a special assistant to the president for consumer affairs can appeal to householders to adopt ways of conserving energy in household heating and cooling. The overall effect of all this pointing out, exhortation, testimony, oversight and coordination in conservation

---

[13]State of New York, Public Service Commission, Joseph C. Swidler, Chairman, *The Energy Crisis—An Overview,* by Alvin Kaufman and Thomas E. Browne, OER Report No. 12, November 6, 1972.

terms is little in comparison with that of the other components of the nation's energy policy. Energy conservation does not now figure significantly in the mix of components which in the aggregate goes for an energy policy.

Even assuming the possibility of mutually consistent policy proposals emanating from the federal bureaucracy, including the chief executive, the congressional obstacle is difficult to surmount. In 1971 the president issued a package proposal of new fuel policies and research, claiming it to be "the first comprehensive energy statement by a president." His proposals for increasing the supplies of energy and for research and development of new sources have already been presented. One conservation measure was proposed: development of a higher standard to increase the amount of insulation required in federally-insured homes. (Heat loss would be lessened by one-third or more.) In the same vein the president earlier had proposed a federal bill for power plant siting; for regulation of sulfur emission and discouraging development of wetlands by taxation; and federal aid to the states in the development of land-use laws. All of these proposals were renewed in a message to Congress a year later.

In view of the known complexity faced by a president of one party trying to obtain enactment of complicated legislation by a Congress controlled by a different party, it is hardly necessary to add that little of this package was enacted into law in the 92d Congress. Mr. Nixon avoided the most politically hazardous energy issues, such as oil import quotas, depletion allowances, control of strip mine damage, the development of Alaskan oil, and the expansion of the oil industry into such diverse interests as nuclear fuels and coal mining. He also went too far and too wide in his proposals for political acceptance of them *as a package* by almost any Congress. Either separately or in combination his proposals run counter to much congressional sentiment, raising doubt as to how much and which parts of his package Mr. Nixon really expected Congress to accept. Use of the federal taxing power to abate sulfur emission into the atmosphere was particularly unpopular. More than half of the eighteen million tons of sulfur spewed annually into the air comes from electricity-generating plants. Yet no supporters of a tax on sulfur could be found among Republicans in Congress to introduce an administration bill. In 1972 it was introduced by three environmentally-minded Democrats.[14]

A national energy policy should be part of an integrated national environmental policy. A major contribution of NEPA is that it is based on a new politics of values instead of the traditional politics of interests.[15] Obviously a new environmental policy is in the making. One segment, a large one, is a national energy policy. In the past policy-making in the United States has been, in general, a composite of particular interests, rationalized in terms of the national interest, and largely determined by economic considerations. It is too much to expect that, even with NEPA, an energy policy can be devised without compromise which will readily incorporate environmental and other noneconomic considerations along with economic factors. But NEPA clearly points in the right direction. The policies, procedures, and devices listed in an earlier part of this book must be selectively improved and enriched by new practices and institutions which will capitalize on the values embodied in the National Environmental Policy Act.

---

[14]*The Washington Post*, July 6, 1972.
[15]Caldwell, *op. cit.*, p. 227.

5.
Science-Based Technology for Environmental Improvement

To come to grips with the complex socio-economic-political problem of environmental policy and science-based technology, if not to solve it, the following program is proposed, with the priorities indicated.

Highest Priority—Sensitize and inform society
Priority 2—Reorient scientists, technologists, and corporate managers
Priority 3—Population control
Priority 4—Change old and create new institutions

A MASS MEDIA PROGRAM TO SENSITIZE AND INFORM SOCIETY

The first point of our four-point proposal, the use of propaganda on a nationwide scale to promote environmental policy, is of the utmost importance. Without a receptive public the other three points of our proposal will hardly be attainable. Nor can they be reached without more general acceptance of the economic philosophy of restrained growth. It is readily conceded that to assign a higher priority to one point of the program than to others is arbitrary, particularly when there can be little doubt that all four points should have high priority. Yet, despite this concession, knowing as we do that the success of any public program involving widespread and marked changes in attitude depends on the active support of a number of alert, informed, and involved publics, the creation of a new opinion favoring the environment, consciously, persistently, and diligently constructed and promoted, is crucial.

Here, the size and nature of the job to be done should determine the kind of talent and the volume of resources to be assigned to the task. By any measure the magnitude and momentum of the environmental improvement program is overshadowed by the sheer weight of traditional attitudes toward resources, their use and conservation, and the "use and throw away" philosophy so insidiously promoted by the philosophy of unlimited growth. The successes of environmentalists pale in comparison with the preponderance of opinion favoring traditional attitudes.

Superficially, leaders of the environmental-ecology movement have gained many successes. They are active, urbane, skillful, and talented, and they are taking their campaign to the opposition on many fronts. There are, among them, scientists in government employ, like the late Rachel Carson, Dr. Kelsey of thalidomide fame, and the Surgeon-General of the U. S. Public Health Service. Eminent private exponents of environmentalism include Joshua Ledberg, Barry Commoner, Rene Dubos, Paul Ehrlich and Ansley Cole, also journalists such as Daniel Greenberg and Frank Graham, Jr., and authors like the late James Newman and publishers like Gerard Piel.

In addition to eminent individuals, interest and pressure groups and public interest law firms have grown in number and influence. They are taking the fight for environmental quality to the courts and to the regulatory agencies and the bureaucracy. Lawyers for the Environmental Defense Fund, Inc., initiated the action in the courts in 1969 against the U. S. Army Corps of Engineers which resulted, a year later, on order of the president in stopping construction of the Cross-Florida Barge Canal. With an advisory Scientists' Institute for Public Information the Natural Resources Defense Council, Inc., inaugurated in 1971 a "Project on Clean Air," a long-term national effort to monitor federal and state implementation of the 1970 amendments to the Clean Air Act. Another public interest environmentalist group, Friends of the

Earth, in 1971 started the legal action against the Federal Communications Commission to compel it to review its fairness doctrine's applicability to spot television commercials. This doctrine was adopted forty years ago to assure the airing of all sides of controversial questions. In lobbying campaigns related to court action environmental lawyers and related groups are lobbying legislation through state legislatures, as in Michigan, recognizing the right of individuals to take polluters to court with the burden of proof on the defendant.[1] A temporary coalition of environmentally-oriented groups helped to block congressional approval of funds for the supersonic transport plane, thus illustrating the astuteness of such groups in legislative lobbying. In announcing in 1971 that environmental pressures were a factor in its decision to withdraw from a group committed to exploit Alaskan North Slope oil, and to write off its unamortized lease investment, the Marathon Oil Company (Ohio) conceded another environmentalists' success.[2]

Such successes reflect a growing environmental-ecological constituency across the country, which is neither unified nor coherent, yet is sufficiently widespread and well enough organized to embody a respectable volume of public opinion. Although Environmental Action officials, sponsors with Senator Gaylord Nelson (D., Wisconsin) and Paul McCloskey (R., Cal.) of the first Earth Day (1970), differ on most things about it, except that it got a good press, there is little doubt that concern for the environment was brought to the attention of many Americans who had hitherto been ignorant, apathetic, or hostile to it.[3]

Soaring membership lists of conservation and ecology-oriented organizations are another indication of the upsurge in interest in

[1]Joseph L. Sax, *Defending the Environment: A Strategy for Citizen: Action,* New York: Alfred A. Knopf, 1971.

[2]*The New York Times,* December 11, 1971.

[3]Stephen Cotton, "Earth Day Got a Good Press," *The Washington Post,* April 18, 1971; Denis Hayes, "Yes, There Was an Earth Day," *The Washington Post,* May 2, 1971. Cotton was press director for Environmental Action at the time. He had been recruited by Hayes who was national coordinator.

the environmentalist movement. The Audubon Society, the American Museum of Natural History, and the Sierra Club all reported stunning increases in circulation of their respective publications, in the eighteen-month period from the fall of 1969 to the spring of 1971. A new source of revenue is advertising from the oil companies. The senior editor of *Natural History,* organ of the American Museum of Natural History, is reported as saying that "They're [the oil companies] trying to get the message across that they're friends of the environment."[4] The importance of women and youth in the environmentalist constituency was brought out in a survey of reader opinion of *Natural History*: ". . . when women and the under thirty group rally at environmental barricades, the generation gap disappears . . . The sample is definitely beginning to challenge a number of assumptions of our society and even to question the capacity of American economic and political institutions to solve environmental problems."[5]

An aroused environmentalist public, a sensitive press, and concerned legislators have made household detergents, especially their phosphate and enzyme content, a matter of everyday discussion and political decision and have placed a $1.2 billion industry on the defensive. Arguing that to replace the phosphates in detergents by other chemicals would be a greater potential danger to public health than the phosphates are to rivers and lakes, the Soap and Detergent Association is suing cities and towns to prevent enforcement of local ordinances prohibiting phosphate detergents.[6]

At the same time the industry is fighting a proposal of the Federal Trade Commission to require phosphate warnings on product labels. Housewives, meanwhile, are creating a demand for detergents that do not pollute.

Regardless of the outcome of these lawsuits and administrative battles, the latest round in the political struggle among producers, consumers, and environmentalists has gone to the producers.

---

[4]Quoted from *Newsday* in *The Washington Post,* April 2, 1971.

[5]Virginia Hine and Luther P. Gerlach, "Many Concerned Few Committed," *Natural History,* vol. 79, no. 10, December 1970, pp. 16-17, 76-80.

[6]*The New York Times,* January 31, 1971.

Claiming that the alkaline content of most nonphosphate detergents was more of a public health hazard than phosphate detergents were to the environment, federal officials appear to be convinced, at least for the moment, that housewives should return to phosphate detergents and have so recommended. As to the pollutant effect of phosphates on lakes, rivers, and streams the EPA administrator would have the country adopt a $500 million program for removing phosphates from sewage.

As regards enzyme detergents the health hazard, present and prospective, in the manufacture and use of such products has been pointed out so persuasively by reputable scientists and disseminated so widely by the media that one of the three big manufacturers is eliminating enzymes from its biggest selling detergent, while the other two acknowledge that changes are being made in the formulae of their detergent products.[7]

No industry is more important in selling the country on environmental policy than television. Yet one gets the impression that its potential in this connection is not being realized. What can be done has already been indicated—the "good press" which the first Earth Day received (1970). Two years later, when the Indochina war, presidential trips to mainland China and the Soviet Union, and the forthcoming presidential campaign had pushed environmental policy off the front pages, the full potential of one part of the television industry was realized for a day. At the suggestion of one promoter of the original Earth Day, WOR-TV (New York) carried a TV marathon—twelve hours uninterrupted by commercials. In whole or in part it was carried by some fifty stations across the country.[8]

Thus greater attention to environmental policy from the mass media can be detected. More public service programs with the environment as the theme are on the air. The conversion of a popular entertainer like Arthur Godfrey to the cause alerts nationwide audiences. Enviromentalists are encouraged, too, by news of such cleanups of the environment as that around Lake Washington,

---

[7]*The New York Times,* January 31, February 14, 1971.
[8]*The New York Times,* March 9, 1972.

in Seattle, as well as of the lake itself, particularly when it has been brought about, at least in part, by one of the media, in this case, local television.

On the surface, then, environmentalists have achieved one or two spectacular successes, many more short-term or intermediate goals, and, from time to time, have succeeded in getting the subject of a higher-quality environment discussed among the public generally.

But relative to the size and complexity of the job to be done environmentalists have not been able to do much more than dent popular thinking. While a few special-interest publics concerned with improving the environment have coalesced in the last decade, and while many of their members are concerned and a few committed, the mass of the public generally does not rate achieving a quality environment very high. While some industrialists and corporations are supporting legislation for a cleaner environment, many more are dragging their feet. While some labor leaders and some in the rank and file among the workers are finding an identity of interest with environmentalists, many more are afraid of losing their jobs. While the media can make an environmental wave when Earth Day was a novelty, similar to Lincoln Steffens's claimed ability to make a crime wave at any time by just reporting to his paper every entry on the police blotter, no commercial television network picked up for special treatment the striking theme and dramatic highlights of the final Report of the President's Commission on Population Growth and the American Future. Instead commercial television left that to the publicly-supported Public Broadcast System (PBS). Three years after President Nixon called on Congress and the people to recognize the crisis and eight months after the congressionally-created commission had submitted its final report, for the first time ever a federal goverment report was made the subject of two-hour-long specific treatment by television, on PBS (November 29, 1972).[9]

The urgent need for sensitizing and informing the public about environmental policy can be brought into sharper focus by re-

---

[9] *The Washington Post,* November 29, 1972.

ferring to the problem of convincing workers and their leaders that
environmental improvement need not affect them adversely and
to the difficulties encountered in selling the recycling of materials
to the people at large.

Basically, workers are skeptical if not hostile to environmentalists' arguments because they fear for their jobs if employers
are forced to meet air, water, and other antipollution standards
set by Congress and enforced by the states.

For years before 1972 when EPA began to bear down on
polluters, industry generally and certain industries—mining and
smelting, iron and steel, paper, petroleum, coal and gas, electric
power generating—in particular had been polluting the environment with impunity. Beginning in 1972 EPA and the Justice
Department, in the federal government, and the several states,
began to bear down on polluters, acting under the 1970 Clean Air
Act, with amendments, and other legislation, such as the Refuse
Act of 1899. Enforcement and administrative problems are
enormous, thus affording laggard or uncooperative industrialists
and firms ample opportunity to fight enforcement in the courts
or otherwise.

Among other reasons given is the cost to industry of building
and installing devices for reducing pollutants to meet the standards.
Sometimes, despite the cost, firms act in good faith and try to meet
the deadlines for pollution control. Sometimes, on the other hand,
mostly for economic reasons (to conform would cost too much),
firms will resist official efforts to compel performance. Here, firms
will find among their employees many who accept employers' arguments that to comply will cost too much, would force shutdowns
and layoffs, or require moving the plant elsewhere. Since antipollution laws are associated with environmentalists, workers tend
to hold them accountable for the supposed threat to jobs, hence
tend to be antienvironmentalist. Hence the difficulty of finding
common ground among industrial workers and workers for a better
environment.

With this as with most public issues the situation is not quite
as simple as the above recital would make it appear. When industries and firms are forced to alter traditional methods and prac-

tices, their spokesmen are inclined to scream louder than the facts warrant. In 1971 industry spokesmen told Congress that the costs of adopting the water pollution control bill then pending would be "astronomical," thus implying that they were beyond the capability of industry to reach. However cost estimates made by the CEQ and for EPA and other federal agencies suggest that price increases to absorb increased costs were well within the ability of the industries studied to handle.[10]

Moreover, general statements such as these rarely reflect the situation in any particular industry, in the economy as a whole, or in any particular plant within an industry. Even industry and plant studies are not always revealing, because industry and plant officials cannot be compelled to open their books for public inspection. As a result, facts for environmentalists to sustain their arguments and for workers to rebut them, and vice versa, are hard to come by. Discussion is lowered to the level of argument and polemics, thus generating more heat than light. Thus, industry's arguments are often persuasive with workers, and environmentalists must find other arguments with which to press their case.

Such arguments are available and are being used with effectiveness. Many workers in big polluting industries are victims of in-plant pollution. Workers in any particular plant suffer along with others from pollution generally. Where old or marginal plants are involved and factors other than the cost of antipollution devices are present, to hold environmentalists responsible is wide of the mark. Moreover, if shutdowns occur in such cases and jobs are lost, workers can avail themselves of existing retraining programs. Through their leaders and lobbyists they can pressure Congress to iron out differences in standards which compel some plants to compete disadvantageously with others. Special cases can be

---

[10]Spokesmen for the U.S. Chamber of Commerce and the National Association of Manufacturers in testimony before the House Public Works Committee, quoted in *The New York Times,* December 1, 1971; Third annual report of the CEQ, quoted in *ibid.,* August 8, 1972; and a study made for CEQ, EPA, Department of Commerce by an outside contractor, quoted in *ibid.,* March 13, 1972.

taken care of by more liberal unemployment compensation programs. Enlightened labor leaders, convinced that environmental degradation is a nationwide problem, and that its costs should be borne in part by stockholders as well as by workers, managers, and consumers, can put pollution controls high on their contract negotiation programs. If profit margins will not permit absorption of increased antipollution costs, this allegation should be subject to independent scrutiny. On all such grounds argument can be made which will show that the interests of workers and of environmentalists can be made to converge, and that workers as well as upper and middle class managers, professionals, and so-called elitists have a monetary interest in cleaning up the environment and in promoting higher quality surroundings.[11]

If industrial workers may be recognized as a segment of society presenting special problems for environmentalists, when it comes to conservation and especially recycling the problems are even more formidable. Here the media will have to persuade much of the population of its environmentally-degrading habits and help change them in favor of new ones promoting a quality environment.

The unpromising progress of waste recycling is discouraging. Except for a few pockets of promise to be noted below the country is not in the mood nor does it appear to have the will to mount a nationwide program of waste recycling. It is not a question of technology. Here technology is the potential agent for improvement in America's life-style and in its environment. For the economy as a whole the technical know-how is known. According to the President's Council on Environmental Quality, "We now have the technology to recycle much of the material that is treated as waste and thereby return it to useful purposes. . . ."[12]

In the metals industries, for example, recycling systems with markets have been in operation for generations. In new industries, too, recycling systems are available. Modern pollution control technology includes such equipment and processes as

---

[11]Stewart Udall and Jeff Stansbury, "Selling Ecology to the Hardhats," *The Washington Post*, April 25, 1971.

[12]Quoted in *The New York Times*, January 23, 1972.

disposing of waste from plastics factories by using it to generate electric power.[13] Another example is the use of recycled glass in large quantities ("glassphalt") as a substitute for limestone aggregate as paving material.[14]

In a slowly-increasing number of cities municipal waste is no longer being used as land-fill but is being recycled. One novel conservation and antipollution process is being used by Ecology, Inc., a Brooklyn, N. Y., company which operates the first commercial plant that processes municipal waste. The process is the Varro Conversion System, a patented system for converting unselected and unsorted municipal refuse into valuable products.[15]

Passing from the particular to the general, some ecologists see in nature the model for answering the baffling question, what to do with waste? Use it, they say, over and over and over again.[16]

However, although attitudes toward environmental problems may be changing from the "emotional peak" of Earth Day, 1970, to a lower, more stable level,[17] pervasive factors operate throughout the country and among the people to obstruct the use of this technology. Congress has been informed by an expert in the field: ". . . we have federal policies and national philosophies which serve as economic obstacles to the use of these valuable and needed materials." He continues: "Present industrial habits, government policies, public apathy, prejudicial and discriminatory regulations put the nation on a path of virgin material preference and direct it away from economically viable recycling."[18]

---

[13]Gene Smith, "Power from Waste Disposal," *The New York Times,* March 5, 1972.

[14]*The Washington Post,* October 1, 1972.

[15]Interview with Hunter McPheters, consultant to Ecology, Inc., reported in *Caduceus,* Kappa Sigma International Fraternity Magazine, May 1972, pp. 18-24.

[16]LaMont C. Cole, "What to Do with Waste? Use it Over and Over and Over Again," *The New York Times Magazine,* April 2, 1972.

[17]Philip A. Abelson, "Changing Attitudes Toward Environmental Problems" editorial, *Science,* vol. 172, no. 3983 May 7, 1971.

[18]M. J. Mighdoll, executive vice president of the National Association of Secondary Material Industries, Inc., in testimony before the fiscal policy subcommittee of the Joint Economic Committee, November 1971. *The New York Times,* January 23, 1972.

This testimony is corroborated by the Council on Environmental Quality itself:" . . . Market and other incentives in recent years have tended to work against recycling."[19]

The forces working for recycling are relatively few. They are dispersed, uncoordinated, and lacking leadership and resources. In the survey already quoted, some business interests in four urban areas, together with numerous local action groups, are noted as having taken the initiative in beginning the campaign to recycle the nation's garbage, now costing the country $4.5 billion a year to dispose of. Denver, Los Angeles, Portland, Oregon, and Franklin, Ohio, are the four cities mentioned. In Denver and Los Angeles pilot projects are under way for recycling paperboard milk cartons; in Portland, Oregon, newsprint; while in Franklin, Ohio, a community of 10,000 south of Dayton, the municipality itself, with the help of federal funds, has constructed and is operating a recycling plant for the city's garbage. There are many other areas where local groups are similarly active. But, in the aggregate, as the earlier quotations show clearly, the *pro* recycling forces are minute as compared with the *con* forces. Even with the technology available, this method of promoting environmental betterment has so far not produced very encouraging results.

Slow progress in waste recycling is matched by snaillike progress in promoting energy-conservation measures. Here, too, the problem is not that we do not know what would be required of us. We do know, or rather the knowledge is available, and it should be made a major part of the mass media program for informing and sensitizing society proposed herein.

In a study, "The Potential for Energy Conservation," prepared by an interagency working group convened by the director of the Office of Emergency Preparedness (OEP) in his capacity as coordinator and overseer of the federal oil import program, energy-conservation measures have been formulated which would produce savings in energy equal to 7.3 million barrels of oil per day by 1980. This is about two-thirds of projected oil imports for that year.

---

[19]*Ibid.*

To realize the full potential four major measures would have to be taken:

1. improved insulation in homes
2. adoption of more efficient air-conditioning systems
3. shift of intercity freight from highway to rail, intercity passengers from automobiles to mass transit, and freight consolidation in urban freight movement
4. introduction of more efficient industrial processes and equipment.

When informed of the magnitude of the realizable savings within a decade, OEP Director George A. Lincoln was incredulous. But, he informed a congressional committee, after scrutinizing the group's report and its work methods, he was convinced that its estimated potential savings were feasible by 1980.

In a mixed frame of mind, the country is not in need of the *techniques* of propaganda. It has these in abundance. But with the active environmentalist groups already showing the way, and with the superb facilities of the mass media at hand, dedicated leadership in the public and private sectors is all that is needed.

REORIENT SCIENTISTS, TECHNOLOGISTS, AND CORPORATE MANAGERS

As a community, or subculture, scientists, technologists, and managers are part of the larger community of the nation, hence are the products of its educational system, sharing and reflecting the values of a growth-oriented free enterprise system.

While there are exceptions, the scientific community in general has identified itself with the corporate interests which have made the country great in the sense of economic growth. Obviously, this is the case with scientists and engineers employed by these interests. It is also true, generally, of those in the universities, who engage in both teaching and research. Particularly is it true of those who have grown up with and in the industries of recent origin, like space and atomic energy, pharmaceuticals and man-made fibers, electronics and communications, public relations and advertising, packaging and air transportation, pipelines and energy, also the entertainment industry.

We do not think it is unrealistic to lump scientists, engineers, and corporate managers together in discussing the necessity for reorientation of attitudes toward environmental betterment. Some, probably many, in all three groups will prove to be intractable. But is it too much to hope and expect that many more will view the situation otherwise? No doubt it is difficult for any group to analyze itself, particularly when vested interests are involved. And many will be unready and unwilling to try to be objective in soberly facing the environmental situation. Yet scientists, trained to look for and act on the basis of facts discovered, will surely recognize the gravity of the environmental situation. Engineers, trained to deal with materials with ascertainable characteristics, will recognize the necessity for community restraint. And managers in the corporate community, trained to act on the advice of scientists and engineers, will recognize the need to develop an environmental component in their scale of corporate responsibilities.

As an institution the modern corporation should continue to be used, not only as a successful device for raising capital, for supporting research and development for higher productivity and a higher gross national product, but also as the dominant institution of our time and potentially the most effective pro-environmentalist institution. As such it should acquire a new dimension in its sensitivity to environmental needs.

On the theory that corporations and their stockholders owe a debt to society for past pollution of the environment without cost to them, some industries and parts of industries and some corporate managers already find it possible to incorporate into their managerial philosophy a greater concern for the environment without neglecting their obligations to stockholders. Managers and leaders in the extractive industries, the automotive industry, and other industries who ungrudgingly accept taxes on pollutants, while finding ways to absorb additional costs without passing them on to consumers, are leaders and managers of the kinds of re-oriented industrial corporations envisaged here. This would incorporate a principle, which a lawyer for the Natural Resources Defense Council, Inc., calls internalizing, into their economic philosophy and would be a long step forward in reorienting scientists, technologists, and business leaders.

In a speech not too long ago, one public utility executive, Charles F. Luce, New York Consolidated Edison board chairman, admonished us to adopt a new environmental ethic, accept a somewhat lower standard of living, a willingness to pay higher taxes, and to make sacrifices of conveniences.

Whether the modern corporation as an institution can live up to its potential is still an open question. Since environmental policy came to the surface as a public issue, corporativism has proved relatively impervious to demands to place it on a par with the traditional goals of profits and growth.

Many gestures and some action in this direction have been made. The gestures are by now familiar to readers—self-serving advertising in conservation magazines, subsidizing dramatic entertainment on public television, etc. Viewed in the aggregate the volume of funds used by industry to lessen industrial pollution by installing antipollution devices, running into the hundreds of millions of dollars, is impressive. Viewed in comparison with the tens of billions of dollars raised by the same industries for research and development, growth, and profit-making, however, the performance loses much of its impact. On the horizon of most corporations the goal of living up to their potential as environment-promoting devices does not loom very large.

And if business and industry do not embrace environmentalism, who but goverment is to make them do it? Even reforming the corporations is not enough, according to one observer. A government responding with energy and sensitivity to the major issues of the day is the only answer.[20] And, according to other observers, it is unrealistic to look for imaginative policies which would include environmental policy from corporate managers.[21]

Managers of government corporations, such as the Tennessee

---

[20]Joel F. Henning, "Reforming the Corporations Is Not Enough," *The Washington Post,* August 29, 1971.

[21]Report from a conference on the evolution of business management structures in *The Washington Post,* May 14, 1972.

Valley Authority, who mend their polluting ways by terminating contracts for coal procured by strip mining, commend themselves to us, even if it means revising production targets with consequent changes in rate schedules and production philosophy.

Made in good faith, this suggestion seems ingenuous on examination. Yet it is the kind of criticism of governmental institutions which should be examined.

For what it believes are good and sufficient reasons TVA is "locked into" the rapidly changing demand-and-supply situation for electric energy, particularly in the Tennessee Valley. At the time of the Korean War (1950-53), when the so-called Cold War was heating up, decision-makers in Washington expanded nuclear weapons manufacturing facilities in the Tennessee Valley, relying on TVA to supply the increased demands for electrical energy. To meet this demand TVA began buying coal, since its steam- and hydro-generating capacity was insufficient. Thus, national necessity, TVA says, dictated its massive coal-buying, about half of which is obtained through strip mining contracts. Parenthetically, TVA says, these increased coal purchases helped relieve unemployment in the coal fields caused by declining markets for coal for home heating and railroad operation. TVA claims to be the only large buyer of strip-mined coal which requires land reclamation clauses in its contracts and oversees compliance. Instead of being sued by three conservation groups (the Sierra Club, the Natural Resources Defense Council, and the Environmental Defense Fund) for failure to comply with the National Environmental Policy Act, TVA thinks it might better be supported by them for enforcing reclamation on land strip-mined to fulfill coal-buying contracts and promote wider use of this requirement.[22]

In recounting this story of TVA's strip mining contracts for coal, the writer exposes the innocence of the layman in the highly-

[22]TVA Annual Report, 1971, pp. 74-77; letter to editor of *Natural History,* vol 81, no. 8, October 1972, p. 11, criticizing E. F. Roberts' review of Harry M. Caudill, *My Land Is Dying, ibid.,* June-July 1972; appeal from Harry M. Caudill for financial contribution to Natural Resources Defense Council, Inc., June 8, 1971.

charged atmosphere of environmental policy-making. As a supporter of two prominent environmentalist groups and a tree-farmer himself, he is opposed in principle to strip mining of coal. On the other hand, as a student of modern policy-making he is too aware of the strength of pressure groups and the insidiousness of propaganda not to overlook insider preference and political motivation and campaign contributions as influences, sometimes determinants, in policy. Such was his experience in this instance. In researching material for this book he discovered the wide context, not limited to the continental United States, in which decision-makers in Washington and TVA managers came to their decisions. If there is a moral it is to suspend judgment on enviromental policy until deep scrutiny of a problem has supplied a clear background of facts.

There are some indications that the needed reorientation, both in the private and in the public sectors, is already under way. In the former, managers and technologists, on their own and prodded by government and activist public opinion, are beginning to change their industrial processes so as to include pollution abatement devices and their R&D policies to include and accelerate the development of more effective pollution controls. As a whole the automobile industry is a convert, although a reluctant and still resisting one. Some units of the pulp and paper industry, of the soap and detergent industry, and most of the integrated and some of the independent units of the oil industry are similarly tilting their public relations, if not their R&D and production policies, in the direction of developing new and perfecting existing antipollution devices.

In the electric power service industry, a publicly regulated monopoly, some old plants are being fitted with devices for controlling sulfur dioxide emissions, for particulate abatement, and methods of waste recycling are being developed and applied. In planning and constructing new plants of all kinds—hydro, fossil fuels, and atomic energy—pollution control, conservation, and safety devices are being developed and provided for. Some of the impetus comes from within the industry itself, both from managers and technologists; more, possibly, from outside, that is, from the increasingly influential environmentalist constituency. State, local and federal governments, through enforcement of recently-enacted legislation, such as the Environmental Policy and Clean Air acts, and state legislation and local ordinances adopted in response to external pressure from reoriented scientists and technologists, and, in some cases from bureaucratic pressures, have added impetus to the needed reorientation.

Support for, instead of opposition to, new procedures for siting new power plants, such as the proposed federal law and New York State's certificate of environmental compatability, shows a new environmentally-sensitive attitude on the part of electric power service leaders and managers.

While the electric power service industry has been slow, in comparison with other industries, in adopting research and development programs as a vital component of management philosophy, it was in the vanguard in its realization of the importance of public opinion to the health and welfare of the industry as a whole. Dignifying its propaganda—the management of collective attitudes by the manipulation of significant symbols—by the new term "public relations," the electric power service industry a half century ago pioneered in the then infant industry of public relations by consciously cultivating good consumer relations.[23] Hence, this industry could well occupy a key position in the campaign which is needed to obtain acceptance on a national scale of the four-point program recommended here. A vast propaganda effort, aimed at sensitizing the public, reorienting scientists and technologists, research and development, and adapting old and creating new economic institutions, for sustained and reoriented economic growth will need the public relations industry for its success. The industry which assisted at its birth could, by leading the private sector faster along the path toward environmental betterment, perform a public service of the highest order.

The key position in federal-state relations of state regulatory

---

[23]Edward L. Bernays, *Biography of an Idea: Memoirs of Public Relations Counsel Edward L. Bernays,* New York: Simon & Schuster, 1965, p. 287.

agencies and departments in enforcing air and water pollution abatement legislation has already been referred to. The original practice of including representatives of industry, agriculture, county and municipal governments on these regulatory agencies may have been stopped and reversed. According to a recent survey perhaps twenty states have taken or plan to take steps to either reduce or eliminate such representation.[24] Probably few would maintain that such steps eliminate the possibility of conflict of interests. By the same token, however, few would maintain that such steps would not increase the likelihood for the expression of what might be called the environmentalist point of view. More state air and water regulatory agencies with fewer representatives of the point of view of major polluters would be a step in the direction of public institutions more fully adapted to promote effective air and water pollution abatement.

Adoption of the recommendations of the National Science Board to the president in 1971 would be a big step in the right direction. Its report is entitled *Environmental Science—Challenge for the Seventies.* At the outset the board points out the difference between environmental quality and environmental science. The former is the goal of the recently established governmental agencies CEQ, EPA; the latter, when fully developed, will provide the know-how.

But right here is the problem. Environmental science is a new discipline, and it has not yet evolved to the point where it can even provide us with an understanding of our environment, let alone prescribe remedies.

Moreover, the condition of the environment is a more serious challenge than the one that prevailed at the time of Sputnik (1957) because environmental science is not prepared now to meet this "perceived challenge" as science and technology were a decade and more ago.

To meet this yawning gap in our knowledge the National Science Board wants a federal commitment to expedite the progress

---

[24]Survey conducted by *The New York Times,* reported December 19, 1971, p. L 44.

of environmental science. The requisite program would focus on solving intermediate-scale problems, on continued effort on global problems, and on ensuring the continued vigor of the necessary research and education to provide the specialists and new knowledge required for environmental science. Among other things support of suitable organizational and employment incentives should be accepted by the federal government as part of its responsibility.[25]

Obviously, the suggestions just made would only be the beginning of a proposal. The reorientation of scientists and engineers, like the building of new and philosophically-reoriented institutions in both government and private industry will begin to materialize when the mass education program envisaged in our first point starts to move. Scientists, engineers and managers cannot be re-educated overnight. Nor can new institutions spring into being at the wave of a magician's wand, not even that of magicians in the garb of scientists and technologists. In both points one and two of our program the most that can be hoped for and effected is a speeding up of the process of adaptation of existing institutions to new conditions and to meet new needs.

POPULATION CONTROL

Rather than attempting to control science and technology as such, the first two points of our program adopt an indirect approach. The same is true of our third point, population control. Not only should the direct approach be ruled out on philosophical grounds as alien to the American ethos. It should be ruled out on practical grounds as well. It is doubtful whether a ban on science and technology, even for the laudable purpose of dealing with the environmental crisis, could be enforced. The success sought for will turn on individual choice rather than government fiat.

Strange as it may seem, of the four points in our proposal population control has the best-developed identity and momentum

---

[25]*Environmental Science—Challenge for the Seventies,* National Science Board, Third Annual Report to Congress, 1971.

of them all. In the country as a whole it must be admitted that there is little if any consensus on the urgency of an educational program. Nor does reorienting scientists and engineers have a dynamic constituency countrywide. And an enhanced federal role for R&D is still emerging. But population control, a taboo less than two decades ago in government circles and in many, if not most local communities, has gained a recognition in the meantime which earlier would have been thought impossible. Although still undergoing attack on moral and other grounds, as a movement population control is growing. There are debates among various disciplines as to the best methods. Some are even advocating zero population growth (ZPG). Yet, regardless of such differences, the consensus is growing, too, that the people-resources ratio is out of balance and that the link must be broken between exponential population growth and environmental degradation.

In recent years dramatic evidence has become available of the crucial role of individual choice in controlling population growth. Since 1955 the U.S. Bureau of the Census has published birth expectation statistics. From data collected in 1971 and published in 1972 it appears that the average number of children which wives between 18 and 24 expected to have dropped sharply between 1967 and 1971 from 2.9 to 2.4. For wives in this age group the average between the years 1955 and 1967 fell three-tenths point from 3.2 to 2.9 children. Two implications are pointed out in the Census Bureau report, one, that more women in the youngest childbearing group want fewer children, and, two, that population figures projected into the year 2000 at the lower number of children would mean a smaller population by 25 million—280 million as against 305 million. One can only guess why wives have fewer children. But, regardless of the reasons, the fact is that the choice is one made by the individual.[26]

----

[26]From data appearing in the Bureau of the Census Report, series P-20, no. 232, reported in *The New York Times,* February 17, 1972. In the *Times* the following reasons are quoted from the Census Bureau's report: a higher marrying age; fewer single women at ages 20 to 24; the economy; the environment; the increase in the number of working women; the

More recent data bear out the trends just noted. For the first quarter of 1972 only 15.8 children were born for each thousand Americans. This was the lowest birth rate ever recorded in this country. A more significant figure, the fertility rate, which measures births per thousand women of childbearing years (15 to 44) reached an all-time low, 74.4, down 11 percent from the corresponding period of 1971. Finally, first-quarter 1972 birth figures suggest that the average number of children estimated for women of childbearing age declined to 2.1, thus continuing the decline from 2.9 to 2.4 noted above in average number of children which wives between 18 and 24 expected to have between 1967 and 1971.[27]

Hard data helping to explain this precipitate decline are not readily available, largely because the reasons why women are having fewer children can be reduced to statistics only with great difficulty, if at all. However, sparse data on the operation of liberalized abortion laws suggest that limiting family size is more important than terminating unwanted pregnancies. In 1970 abortion was legalized in New York State. Reporting in 1972 on the reasons for seeking abortions given by 278,122 women who had legal abortions in New York City during the first eighteen months of the law's operation, Gordon Chase, New York City's Health Services Administrator is quoted as saying that "New York City women tended to get abortions to limit family size while non-residents tended to get abortions for a first out-of-wedlock pregnancy.[28]

From the data cited it would appear that, in some circles at least, some young women the country over, and some young women in New York City, are taking population control seriously.

---

women's liberation movement; the "marriage squeeze," that is fewer older men (those born in depression and war years) and more young women (who were born during the postwar baby boom). It is generally known, too, that improved contraception and liberalized abortion laws are important in this connection.

[27]Data from National Center for Health Statistics reported in *The National Observer,* week ending June 17, 1972.

[28]*The New York Times,* February 20, 1972, p. 41.

Thus, even before the findings and recommendations of the Commission on Population Growth and the American Future were published in 1972, the trend toward the Commission's goal of a gradually stabilized population had already been established. But, since trends characteristically change without explanation, the Commission recommended that policies already adopted by the federal government be extended and expanded to maximize what the Commission believed to be the steps already taken explaining the trend toward a stable population. These are elimination of poverty, stamping out racism, giving women equality with men, and enhancing the freedom of choice of all Americans to avoid unwanted births and realize their own preferences in childbearing and family size.

Not all the Commission's recommendations were unanimous. Opinion was divided on expanded sex education and the sale of contraceptives to minors. And President Nixon rejected the Commission's recommendation for legalized abortion. It was, he said, an unacceptable method of population control.

But the Commission found unanimously that it was desirable to bring U.S. population growth to an end. The concept of population stability rather than continued growth is the heart of the Commission's conclusion.

On the relation between population growth and the economy the Commission's findings were negative: ". . . We have looked for, and have not found, any convincing economic argument for continued national population growth. The health of our economy does not depend on it, nor does the prosperity of business or the welfare of the average person."[29]

But the 1972 Stockholm Conference on the Human Environment poses a dilemma for the American people and their political leaders. Two-thirds or more of the 113 countries represented there are the less-developed countries. A stabilized population in the

---

[29]*Population and the American Future,* the Report of the Commission on Population Growth and the American Future, Washington, D.C.: U.S. Government Printing Office, 1972.

foreseeable future, a slowing down of economic growth, the fixing of priorities emphasizing quality rather than quantity, particularly in environmental matters: these are policies suitable for a developed country like the United States. But the peoples and leaders of the less-developed countries view population trends and economic growth and pollution and other environmental problems in different terms.

The growth rate may be cited as an example of the difference in approach. Whereas in the United States there is a growing conviction, emphasized by the Commission on Population Growth, that a continuously expanding economy exacerbates environmental degradation and hence is to be questioned, in the less-developed countries almost without exception the foremost national goal is accelerated economic growth, with consequences to the environment receiving little or no attention. In a conference such as that at Stockholm in 1972 much of the dialogue in the United States over economic growth seems wide of the mark. The dilemma for the United States, as for all developed countries, can be stated in terms of priorities. Given finite resources, with an expanding world population, which things come first, the necessities of life for all human beings or affluence for some? The dilemma should be thought of in terms of challenge: organization of science-based technology so as to maximize the possibilities of a decent environment for all mankind.

The recommendations of the Stockholm Conference on the Human Environment were adopted by the United Nations General Assembly at its twenty-seventh session (1972). Resolutions were adopted to establish an Environmental Secretariat and a Governing Council. One of the Council's purposes is to coordinate the work of the U.N. Specialized Agencies (such as the World Health Organization, the International Labor Organization, the Food and Agriculture Organization, etc). Since most of the specialized agencies have their headquarters in Geneva (Switzerland) some delegates thought the new agency's headquarters should be in Geneva. On the other hand, more delegates, mostly from the Asiatic and African states, wished the headquarters to be in Asia or Africa, since, among other reasons, most of the agitation for slowing down population and economic growth has come from the Western states, thus raising suspicions among the Afro-Asian bloc. The result was the selection of Nairobi, in Kenya, as the seat of the new agency.[30]

Interest, pressure, propaganda and other citizens' groups working on the domestic scene are now joined, as a result of the 1972 Stockholm Conference on the Human Environment, by the undeveloped nations lobby. By definition all members of this unofficial yet influential congeries of nations are less developed than the developed nations; that is, the former are at a less advanced stage of economic development than the latter, hence are eager to accelerate the rate of growth of their national product. Yet it is this very obsession with increasing economic growth which, in the developed countries, is one if not the main cause of environmental degradation. This poses a dilemma for existing intergovernmental organizations, such as the United Nations, and the North Atlantic Treaty Organization (NATO), as well as for the new ones resulting from the Stockholm conference. In addition, national governments and particularly that of the United States will have to reexamine their philosophies of foreign aid. In effect this means that every interest, pressure, propaganda and other citizen group interested in promoting environmental improvement in the United States will perforce be confronted by this new dimension of environmental problems viewed worldwide. As seat of the United Nations, New York will share the resources and attention of these

---

[30]Popular among many, although not all, environmental advocates in the Western world is the pseudo-scientific work. *The Limits to Growth: A Report for the Club of Rome's Project on the Predicament of Mankind* (New York: Universe Books, 1972). The thesis of the authors is simple: "If the present growth trends . . . continue unchanged, the limits of growth on this planet will be reached sometime within the next hundred years." By definition the less-developed countries have lower per capita incomes than the developed, mostly Western, countries. Emanating from the West, the argument that both population and economic growth should be limited is not popular among the less-developed countries. Hence the suspicion of the West.

groups along with Washington as the seat of the United States government.[31]

CHANGE OLD AND CREATE NEW INSTITUTIONS

In proposing to change old and create new institutions, the fourth part of our proposal concentrates on the establishment of a new federal office of antipollution scientific research and development modeled on the World War II Office of Scientific Research and Development (OSRD) but devoted to the development of environmental science. Together with the other three parts of our proposal, particularly the reorientation of scientists, technologists, and corporate managers toward environmental policy a new federal office dedicated to environmental science could be the key institution by and through which each of the other parts of our priorities program could reinforce each other.

With such a federal office in being the mass media would have a clearing house for information on the current progress being made in environmental science, thus providing dramatic new material on developments in science fully worthy of regular and systematic coverage.

With such a federal office in being the place of the component parts of environmental science would be upgraded in the scientific hierarchy, thus stimulating education and training of young scientists and engineers and drawing them to an institution created expressly for the most efficient use and best employment of environmental scientists.

With such a federal office in being the existing federal Bureau of the Census and other statistical services would discover and develop mutually rewarding contacts in the field of demography, particularly population control.

And with such a federal office the institution itself would take pride in knowing that its reason for being was to accelerate the

---

[31]See David A. Kay and Eugene B. Skolnikoff (eds.) *World Eco-Crisis, International Organizations in Response,* Madison: University of Wisconsin Press, 1972; also *The World Bank and the World Environment.* Washington, D.C.: World Bank Group, September 1971.

development of environmental science the better and the more rapidly to improve man's habitat.

Through the president and Congress the National Science Board has told America that environmental science is not as well prepared to meet today's vast problems of environmental degradation as space science was prepared to meet the 1957 crisis precipitated by Sputnik.

In today's crisis we need to recapture the spirit which animated government and people during World War II, particularly the partnership between the military, the scientists and the engineers, including those from the universities, in inventing and developing new weapons of war.

From the scientific and technological points of view the fantastic success of the United States and its allies in creating new weapons of war—the atomic bomb, improved radar, and the proximity fuse, to mention only three—is due to the institutionalization of this partnership in the Office of Scientific Research and Development (OSRD). Enjoying the complete support of the president and commander-in-chief, Franklin D. Roosevelt, and operating under the leadership of an able scientific administrator, Vannevar Bush, who was at the same time the president's scientific advisor, OSRD accomplished miracles during the war. But it did not survive into the postwar period. Some of its component parts did survive, however, and they or their counterparts continue the tradition of civil and military "partners with mutual respect" which OSRD fostered during the war, at least so far as research and development (R&D) on new weapons and weapons systems are concerned.

Early in the nineteenth century technology and engineering, particularly as stimulated by defense and security requirements, bred more science-based technology, just as Woodrow Wilson, a half century later, found that "legislation begets legislation." In the post-World War II period of this partnership between military and civilians certain components of civilian industry and the military became so close that the term "military-industrial complex" was coined to describe it.

With the fruits of scientific research so widespread both American communities, scientific and lay, are interested in technological innovations. Hence the history of the wartime OSRD is instructive.[32] For them and for others concerned with the environmental crisis the memoirs of OSRD's director are not only instructive but also suggestive.[33] Granted that America's participation in World War II and the current environment crisis are not only different but are of different orders of magnitude, nevertheless, the similarities are obvious enough to warrant scrutiny of the idea of another OSRD reoriented to researching and development of the technological needs of a massive environmental improvement program.[34]

OSRD was operated, if it was not set up, to meet Bush's theoretical model, i.e., a loose structural organization for rapid progress on weapons in close collaboration with a closely-knit military organization, the supreme command overseeing the whole operation from a distance but with the justified assumption that this collaboration was cordial.[35]

A revival of the OSRD idea, suitably adapted, would replace the military component by civilian, industrial, manufacturing, mining and probably other components.

---

[32]See Irvin Stewart, *Organizing Scientific Research for War*, Boston: Little, Brown & Co., 1948.

[33]Vannevar Bush, *Pieces of the Action*, New York: William Morrow and Company, Inc., 1970.

[34]Worth noting is the organizational form of the World War I National Defense Research Committee (NDRC), which was continued into OSRD. At NDRC's first meeting a pyramidal form was decided upon. That is, each independent member (not representing the NAS, the army, or the navy) took charge of a divisional operation by delegation from the committee. Thus, to Arthur Compton was assigned radar; to James B. Conant, chemistry and explosives; to Frank Jewett, communications and transportation; to Coe, patents and inventions. Vannevar Bush was the chairman. The theory was that to each division head there would be a large delegation of authority, including authority to set up sections and to recruit specialists to man them, with programs developed in the sections moving up. In a separate office reporting to the chairman all business and government relations were centralized. There was, too, a separate office of contracts with educational institutions financed by NDRC. Later, an office of field service was set up. (Vannevar Bush, *op. cit.*, Chapter II. "On Institutions.")

[35]Bush, *op. cit.*, p. 30.

OSRD features which appear to be applicable to the present environmental crisis include:

1. high quality leadership
2. recruitment of similarly high quality professionals—scientists and technologists from civilian life, particularly the academic community—thereby facilitating its method of operation
3. a two-way flow of ideas defining and researching promising concepts; originating both within the functional sections and moving upward and at the top moving downward for professional scrutiny
4. semiautonomous status in the government
5. not restricted by either patent or profit considerations

Establishment of a new office of environmental nonpollution research and development embodying these features of the wartime OSRD would promise quick and coordinated action to meet the technological needs of an expanded and accelerated R&D program for a quality environment. Leadership would be upgraded and centralized. High quality professionals could be recruited by the federal government from civilian life, academic and industrial, to work with equally high quality professionals in the government both civilian and military, to identify the weak spots in our present R&D program and move to fill them. The modus operandi which proved so successful in OSRD in wartime would be adopted and, if necessary, adapted to the peacetime war against environmental degradation. A semiautonomous status in the federal bureaucracy would imply freedom to move quickly in meeting needs, unencumbered by bureaucratic rivalries and infighting. If patent and profit considerations stood in the way of quick action public service patents and modified cost-plus contracts could be used. Existing institutions, such as NAS-NRC, NAE, CEQ, and OST, need not lose their identity, and could be fitted into a federal structure giving antipollution research and engineering a high priority status, as could such other outstanding agencies as NIH, NSF, and military R&D. If the difficulties and problems appear great it is because they are great. But the prize to be won is even greater—a wholesome, high quality environment for the American people achieved through full use of America's superior resources in research and development.

Establishment of such an office, moreover, would mobilize public and private research and development for environmental improvement in a way and on a scale unattainable by any other device. It is no reflection on the National Academy of Science-National Research Council, the National Academy of Engineering nor on the Office of Science and Technology; on the office of the president's science advisor; on the National Science Foundation and on the other organizational devices, such as the Council on Environmental Quality, to argue, as we do (on pages 48-52), that such structural changes and organizational reshufflings fail to recognize the gravity of the environmental crisis. All politics aside, if we already have the technology for halting environmental degradation, as is contended in some industrial, governmental and other quarters, a crash program should be mounted to obtain the necessary use of this technology wherever pollution continues. If, as we suspect, this estimate of our antipollution technology is overoptimistic, a crash program comparable to OSRD's wartime orientation to new weapons, but directed to search out the gaps in our knowledge of antipollution technology, should have high priority. A peacetime counterpart, properly financed and staffed, would amount to public recognition by the Congress of the important place which public and private R&D should occupy in the rank order of things. It is not unlikely that certain research establishments, such as the National Institutes of Health (NIH), should be left autonomous or with a special relationship to the reshaped OSRD. There would be numerous, possibly many, such special relationships to be worked out. The important thing, however, is to get acceptance now, before the environmental crisis worsens, of the principle of unity among all research and development efforts, both public and private, in antipollution technology.

An indication of the kinds of problems which a new federal Office of Antipollution Scientific Research and Development might undertake is contained in a suggested list of projects, appearing at the end of this book.

Suppose we look at the electric power industry for an illustration of the need for a public agency or institution designed to provide guidance, fix priorities, and supervise research and development projects to promote environmental science generally and reduce air pollution in particular.

Over a period of years desulfurization of smokestack gases from steam power plants has been universally recognized as one of the obvious ways to abate pollution of the air. Despite the initiation of at least thirty-seven projects involving the expenditure or commitment of over $250 million of government and private funds no substantial progress has been made. Pollution from this source the country over is as bad as ever. Obviously, lack of money is not the explanation. What was lacking, according to Joseph C. Swidler, was a research commitment, effective central direction, a willingness to change outworn attitudes, a continuous search for the most productive approaches, a total immersion in the world of science and research under the best available scientific leadership, and an organization which could bring order out of chaos in electric power research planning.[36]

How should the proposed OAPSRD be financed? Should federal or private funds have priority?

One step toward answering this crucial question has been taken by the electric power service industry. In setting up, in 1972, an Electric Power Research Institute (EPRI) the industry had before it an estimate of the cost of a long-term national research commitment for the industry amounting to $32.5 billion. To the end of the century, this would average out at something over $1.1 billion per year. If this industry, albeit the nation's largest, requires funds for R&D of this order of magnitude, one hesitates to guess how much industry as a whole needs. Whatever it is, it is almost certain to be more than industry as a whole or any single industry

---

[36]Joseph C. Swidler, "The Public Stake in Energy R&D," a talk to the Edison Electric Institute, June 7, 1972. Mr. Swidler was chairman of the Federal Power Commission in 1963 when he made the suggestion eventuating nine years later in EPRI. He is now (1973) chairman of the New York State Public Service Commission.

can or will mobilize. Whether or not the electric power service industry is able to raise all the funds required, it expects a federally-funded program of equal size to be mounted and maintained.

Although Joseph C. Swidler was among the first to advocate a federal tax on energy use as a means of financing electric power service industry research, he did not suggest what the rate should be. A rate of 0.15 mills per kwh on all energy generated throughout the country has been suggested by Senator Warren Magnuson (D., Wash.). But this would produce, at present levels of generation, only $300 million, as compared with needs, as already indicated, of funds almost four times as much. Clearly, if the federal government, through such a tax at this rate, were to raise only a quarter of the funds per year which the electric power service industry needs to solve the most urgent research problems facing it(control of air pollution, dissipation or use of waste heat without damage to marine life, and protection against radiological hazards), private financing would have to meet the other three-quarters in addition to funds needed for research and development to meet the environmental improvement programs of all the rest of industry.

Funds of what order of magnitude are we discussing? To return to the electric power industry again, here the average research expenditure is less than one-quarter of one percent of revenues, an admittedly inadequate expenditure, particularly in view of the willingness of some state public utility regulatory commissions to approve as much as 2½ percent of annual revenues for research purposes. With a gross national product of a trillion dollars at the lower percentage we are talking of funds in the neighborhood of $2.5 billion; at the higher percentage, in the neighborhood of $25 billion. If we were to adjust these rough figures downward by a factor of five we would still be talking of funds in the neighborhood of between $0.5 and $5 billion, far more than could be raised by any federal tax on energy use proposed thus far.

Although the research situation in the electric power industry is not here suggested as a model for the country's industries, still this exercise should be helpful in considering whether a national research and development program as envisaged for the proposed

OAPSRD should be financed by the federal government and private industry jointly. It should be helpful, too, because, as Commissioner Swidler has said, technological improvement in the power industry is more than an industry objective. It is a national goal of high priority, including the tackling not only of urgent problems already listed, but also intermediate problems. Such problems are (1) developing processes enabling us to use abundant coal resources in place of imported oil; (2) improvement in the economy and reliability of generating sources; (3) conservation of fuels and power capacity by avoidance of waste and by obtaining greater efficiency in producing and using energy; (4) improvement in transmission technology which will lessen the drain on land and scenic resources; and, finally (5), a demonstration from the electric power industry of dedication to discovering energy systems, whether based on solar, fusion or hydrogen sources, or some other technology, which can produce electricity without environmental deterioration or exhaustion of depletable resources.

To reinforce the plea made here for new and environmentally more sensitive institutions, public and private, in a context of population control, environmentally-oriented scientists and technologists, and a mass media program to inform and sensitize the public to environmentalism, let us add an example of what happens in the absence of an effective coordinating and supervising mechanism for public and private research and development.

Science-based technology has provided us with a fantastic new tool for discovering new resources and observing the condition of known resources. Using the combination of remote sensors and high-resolution cameras mounted on earth satellites, the Earth Resources Observation Satellite (EROS) has already expanded our knowledge of the resources of the sea; of the whereabouts of new mineral deposits; of the existence and prevalence of disease in crops and in forest resources; as well as helping archeologists in locating and tracing prehistoric ruins. With further experience EROS, a highly sophisticated tool, promises even more surprising benefits to society.

Announced in 1966, with a first launching scheduled for 1969, it was not until 1972 that ERTS-A (for first Earth Resources Technology Satellite) was put into orbit. An explanation of the delay can perhaps be found in the rivalry among departments and agencies over control of the project, in the higher priority given sensors for military purposes (the Department of Defense "spy in the sky" program), and in NASA's Apollo program. In any case, the relatively small resources in funds and manpower devoted to the program combined to make it a "stepchild" of the Apollo program, to use words of Representative Larry Winn (R., Kan.). Lacking a quarterback, yet supposedly supervised by the President's National Aeronautics and Space Council, it was found in a conference in 1972 that the federal agencies had not been speaking to each other; that each was trying to get the lion's share of the larger appropriations expected with the winding-down of the Apollo program; and that new, high-resolution cameras developed for the air force had been withheld from the civilian satellite program. Meanwhile, interested bureaus in the Department of Commerce were shifted in a large-scale government reorganization which brought more administrations into being like EPA, which were also interested in the earth satellite program.

Not all of the delays, frustrations, friction, and rivalries can be attributed to the absence of a mechanism such as is proposed in OAPSRD. Even in the smoothest running bureaucracy all such obstacles to progress are "par for the course." Yet one not unfamiliar with bureaucratic and corporation politics cannot help wondering whether a "federal mechanism" such as is proposed here (Office of Antipollution Scientific Research and Development), operating with the priority for civilian environmentally-oriented projects inherent in such a mechanism, could not have expedited EROS through the maze of federal corridors. When the executive secretary of the President's National Aeronautics and Space Council predicted at the conference mentioned above that the management problems should be solved in a couple of years (from 1972), one participant is reported to have exclaimed that they should have been settled two years ago, if not earlier. Apparently, for whatever reasons, there was a good deal of slippage in the program.[37]

---

[37]*The New York Times,* February 4, 1972.

In accelerating our efforts to harness science-based technology to environmental science and through it to environmental policy we would do well to bear in mind the wise and prudent note struck by a New York business man, Kurt Salmon:

> *"Quidque agis prudenter agas*
> *et respice finem—*
> Whatever thou doest, do it
> well and think
> of the consequences."[38]

---

[38]"Man in Business," *The New York Times,* March 5, 1972.

Appendix

Note on priorities: Appearance of an item at any place in the following list is not to be considered the standing which should be accorded the item.

1. production of electricity from geothermal energy, including the following but not limited to them: environmental problems such as smell and noise of rushing steam; corrosion from mineral salts in hot brines; drilling techniques; finding methods to use hot dry rock zones; separation of minerals; exploration work; development of desalting methods

2. production of electricity from solar energy

3. magnetohydrodynamics

4. production of electrical energy from nuclear fusion

5. increased efficiency of existing methods of burning hydrocarbons, e.g., the internal combustion engine

6. crash program for multiple use of materials—"use virgin materials over and over and over again"

7. crash program for recovery of oil from oil shales and tar sands

8. power plant siting problems requiring priority R&D: waste treatment and disposal; waste treatment rejection; underground transmission; advanced siting practices

9. safety standards for nuclear reactors: emergency core cooling systems (ECCS); safety standards for radioactive emissions

10. disposition of spent radioactive materials from nuclear reactors

111

11. improved efficiency for residential and office building heating and cooling

12. review of and evaluation of tax components of a national energy policy

13. R&D on substituting mass transit for individual motorized transport

14. study of and recommendation on procedures for reconciling different views on disputed scientific matters

15. study of and recommendations on new institutions, public and private, for assimilating science-based technology into public policy, environmental and other

16. research on energy conservation methods

17. research on technological assessment, especially the idea of an independent mechanism, such as the General Accounting Office

18. research and development on agricultural animal waste management

19. research on the essential economic, political, and administrative arrangements and institutions for applying the results of environmental science, as recommended by the National Science Board

Index